essential atlas
atlas

of **botany**

BARRON'S

First English-language edition for the United States, Canada, its territories and possessions published in 2004 by Barron's Educational Series, Inc.
English-language edition © Copyright 2004 by Barron's Educational Series, Inc.

Original title of this book in Spanish: *Atlas Básico de Botánica*
© Copyright 2002 by Parramón Ediciones, S.A., World Rights
Published by Parramón Ediciones, S.A., Barcelona, Spain

Authors: Parramón Editorial Team
Illustrations: Archivo Parramón, Farrés Il·lustració, A. Martínez
J. Torres, Studio Cámara

Text: Josep Cuerda

Translation from Spanish: Eric A. Bye

All inquiries should be addressed to:
Barron's Educational Series, Inc.
250 Wireless Boulevard
Hauppauge, NY 11788
http://www.barronseduc.com

International Standard Book Number 0-7641-2709-8

Library of Congress Catalog Card Number 2003109122

Printed in Spain
9 8 7 6 5 4 3 2 1

FOREWORD

This atlas of botany makes it possible to get acquainted with the plant world, from plants that are invisible to the naked eye, such as microscopic algae, to the gigantic trees that make up the dense forests of the warm, humid regions of our planet. It is therefore a very useful instrument in helping us learn about the great variety of shapes and styles of plant life that can be found in different climates and soils.

The various sections that make up this work are a true compendium of botany. The sections consist of many pages and illustrations that are scientifically accurate even though they are schematic; they show the main characteristics of the anatomy, physiology, and reproduction of the different plant groups and species. The illustrations, which are the heart of this volume, are complemented by brief explanations and notes that facilitate understanding of the main concepts, as well as an alphabetical subject index that makes it easy to locate any topic of interest.

In undertaking this atlas of botany, our purpose was to create a practical and instructional work that would be useful and accessible, highly accurate scientifically, and enjoyable and clear. We hope the readers agree that we have achieved our goals.

CONTENTS

INTRODUCTION

We know the reproductive bodies of certain species of fungi as mushrooms. Some mushrooms are edible, and others are very poisonous.

BOTANY

Botany is the branch of science that deals with the study of **plants**. Long before humans understood how these organisms work and the causes of the great diversity of plant forms that exist in nature, they were deeply interested in these living creatures that are so different from animals.

Primitive humans knew plants for their **usefulness**. Some were edible and provided food, and others were harmful and even poisonous. Wood, stems, and leaves from certain plants were used for making huts, boats, hunting and fishing implements, baskets, and a great many utensils. Certain plants had healing powers or numbed pain; others increased physical and mental abilities. One of the most transcendent steps in the history of humankind was the **domestication** of certain plants and the development of agriculture. From merely collecting wild plants, humans advanced to protecting the ones they valued most and collected their seeds for planting. That is how the road to selection began and

led to the plants that farmers all over the world now cultivate.

Humans have never ceased to be interested in plants. In ancient times, some three thousand years ago, people began to organize their knowledge about them. However, for a long time, botany was nothing more than a branch of medicine, and it was not an independent science until the sixteenth century. A century later, the discovery of the **microscope** allowed botanists to see structures and details of plants that had previously been invisible to the naked eye. From that time on, the general knowledge of humans grew by leaps and bounds.

THE CLASSIFICATION OF PLANTS

In studying plants, botanists have always felt the need to order and classify them in groups with common characteristics. The first classifications were based mainly on the exterior appearance of the organisms, in other words, on their **morphology**. Later on, with the development of theories on **evolution** of life on our planet, all living creatures were classified according to the degree of **similarity** among them. That way, all plants that make up a single group have descended from the same ancestral species that evolved toward different life forms that are better adapted to new situations.

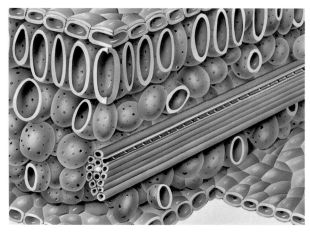

With the aid of a microscope, we can see the tiniest structures of plants.

A tree is a living plant at least 15 feet (5 m) tall, not counting the roots, which secure it to the ground.

THE FIVE KINGDOMS OF THE LIVING WORLD

In Antiquity, living beings were classified into two large kingdoms: plant and animal. Nowadays we don't feel that this **classification** corresponds to the genealogical tree of living beings, and living beings are now classified into five kingdoms. The most basic of them is that of the **Monerans**, which are the bacteria and the blue-green algae. All other one-celled creatures, along with the multicellular ones that lack tissues and differentiated organs, make up the **Protista** kingdom, which includes the algae (except for the blue-green algae) and the protozoa (which previously had been considered one-celled animals). The **Fungi** make up a separate kingdom; they are considered to be plants because they live attached to the soil or other substrate, but they have no chlorophyll and carry out no photosynthesis, in contrast to the algae and plants. The **Plant** kingdom is composed of the terrestrial plants that have chlorophyll—the green plants that we see. The fifth kingdom is that of the **Animals.**

However, we still designate as **plants** all organisms that have a rigid wall surrounding their cells and that are incapable of moving by themselves, the way animals and protozoa do. Similarly, **algae** are the plants that have chlorophyll, live in water, and have a simple body with no reproductive organs, conductive vessels, or special tissues for sustaining themselves.

PLANT ANATOMY

Anatomy is the branch of biology that focuses on what living creatures are like inside. Quite a number of a plant's characteristics can be seen at a glance, as long as we are dealing with a multicelled plant of a certain size. For example, you can pull the petals off a rose and count them, and see how the stamens are arranged inside the corolla. But to observe and clearly distinguish the different parts that make up each stamen, you need a magnifying glass.

The simplest plants are **unicellular**, that is, their body consists of a single cell, as with many algae and fungi. Other plants have a body composed of many cells that are interconnected. The internal structure of **multicellular** beings can be very complex. Different types of cells cluster together to form **tissues** with specific functions. They, in turn, are organized in more complicated functional structures known as **organs**. Finally, a set of tissues and organs working together in a coordinated way constitute a **system**, which carries out a certain series of vital functions, such as the flower, which is the reproductive apparatus of higher plants.

PLANT PHYSIOLOGY

Physiology is the branch of biology that focuses on how living beings function: how they get nourishment and breathe, how they grow, how they protect themselves against unfavorable conditions and their enemies, how they relate to one another,

Beets are a very useful cultivated plant. The red beet (right) can be eaten raw or cooked; sugar beets are used to produce sugar; and fodder beets are used as animal food.

how they reproduce, and so forth. Because every plant species functions differently, there can be plants in all parts of the world, and there can be a great **diversity of species** in a single place.

A tree from the tropical rain forest couldn't live in a cold climate. Its leaves, which stay on all year long and have no protection against cold temperatures, would freeze in the winter. But there are other types of trees that are adapted to functioning in these conditions, such as firs and birches, just as desert plants have special mechanisms to withstand the long droughts and strong sun that plants from other areas couldn't tolerate.

In the same way, within a mass of vegetation, species with superficial roots live with others that have deep roots that feed from different levels in the subsoil. There also are plants that use the trunks of other plants for support as they incline toward the sun, thereby avoiding the need to grow their own trunk, as with climbing vines. Flowering time is another of the physiological traits that help with coexistence, because, by flowering at different times, the plants don't all have the same needs at once, but rather these needs are scattered throughout the year.

REPRODUCTION AND HEREDITY

If individuals didn't leave **descendents** before dying, life on our planet would disappear. One of the main characteristics of living creatures is, in fact, the ability to reproduce, in other words, the formation of a new generation of descendents similar to the **progenitors**. The transmission of traits and characteristics from the parents to the descendents, known as **heredity**, takes place through the **genes**, which are located in the **chromosomes** inside the cells of every living creature.

The type of reproduction used by pine trees and all plants that produce seeds, as well as many other plants, is called **sexual reproduction**, because the embryo is formed by the union of two germinating cells of different sex: one male and the other female. This is the same reproductive system that humans and higher animals use. But the vast majority of plants also can reproduce **asexually**, whether through spores, parts of their body, or other such systems. Humans have used this property of plants from time immemorial for agricultural purposes by practicing propagation through **cuttings** and **grafting**.

A pinecone is a strobile or cone (a composite fruit made up of bracts and seeds) from a pine tree. The seeds are pine nuts, which are edible.

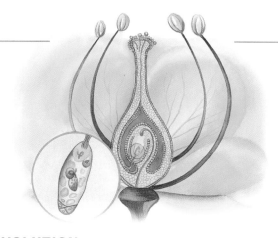

The flowers of certain plants contain a set of reproductive organs.

EVOLUTION

By means of sexual reproduction, there is **variability** among the descendents because of the mixture of paternal and maternal genes. This is clearly visible in human families: the children resemble one another, but they are not exactly the same, and some resemble the father more, while others look more like the mother. This variability, along with other accidental changes known as **mutations** that can occur in the genetic material, means that whenever there are changes in environmental conditions, there are some individuals that are more likely than others to survive in the new situation.

Living creatures make up **populations**. A population is a set of individuals from the same **species** that live in a certain place and relate to one another and reproduce. Because of variability, the individuals of the population that are best adapted are more successful and produce more offspring than the less successfully adapted ones. The latter can't compete and eventually disappear. This is the process of **natural selection**, which has been going on in the living world ever since the beginning of life on our planet.

The primitive plants were one-celled algae that had evolved for millions of years toward increasingly complex multicelled forms, which are the algae that we see on the bottom of the ocean at the shore. Some of these algae evolved toward forms with new characteristics that allowed them to adapt to life outside the water, although in moist locations, as with mosses and ferns. But evolution is a process that doesn't stop, and new species of plants increasingly suited to life on terra firma and very diverse conditions of moisture and dryness kept cropping up.

The passion flower or *Passiflora* was named because people think they can see in it the crown of thorns, the nails, and the hammer that were used in the crucifixion of Christ.

ECOLOGY

Ecology is based on the fact that nothing works in isolation in nature, and that all things and all living creatures are interrelated in such a way that a **balance** is maintained. Any change in physical or biological conditions upsets these relationships and can mean the disappearance of certain species and whether or not new ones appear. Plants are the basis for all other life forms. If there were no plants, there would be no herbivorous animals and, therefore, no carnivores. Naturally, we humans also would not exist. That's the ecological importance that plants have and why we need to protect them.

In every part of the world, there are certain plant species that can live in the prevailing climatic and soil conditions. But we find only the ones that have triumphed in competition with their neighbors that have similar needs. That is the **plant community** of that location, which shares the space with the **animal community**. Both communities, along with the **physical surroundings** and the prevailing physical environment (climate, soil, and so forth), make up an **ecosystem** in which all factors and components influence and interact with one another.

CELLS, TISSUES, AND ORGANS

All living beings are made up of cells, which are the smallest existing units that have their own life. We can see them only with the aid of a **microscope**. The organisms that we can't see with the naked eye are made up of a single cell; that is, they are **unicellular**. The plants and animals that you commonly see are **multicellular**: they have **tissues** composed of many cells, closely associated with one another, that carry out the same task, as well as **organs**, made up of various tissues that perform an important life function in a coordinated way.

PLANT CELLS

All cells basically consist of a viscous liquid known as **cytoplasm** surrounded by a covering called the **cellular membrane**. The **cellular organs** are immersed in the cytoplasm; the most important of these is the **nucleus**, which acts like the brain of the cell. The nucleus contains the **chromosomes**, which carry the genetic information. Other important tiny organs include the **mitochondria**, which produce the cell's vital energy, and the **endoplasmic reticulum**, in the walls of which are the **ribosomes** where **proteins** are manufactured. Plant cells also have **plastids** that contain pigments, **vacuoles**, or cavities, filled with water containing food substances, and a rigid **cell wall** that encircles the membrane.

WITHOUT WALLS

One-celled fungi live in the damp covering of dead leaves on forest floors; they are the only plants that are capable of changing shape, since their cell lacks a rigid cell wall. These are the **slime molds**.

Algae and **fungi** don't have real tissues. **Mosses** and other plants that live in damp locations have very simple tissues. The plants that need tissues and organs are the ones that live on terra firma and are not bathed with water, in other words, the true **terrestrial plants**.

A TYPICAL PLANT CELL

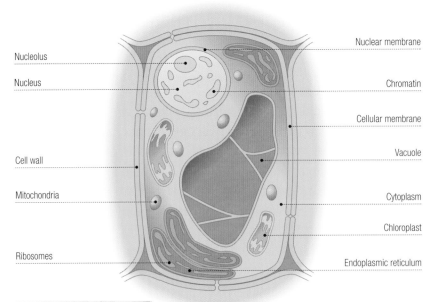

- Nucleolus
- Nucleus
- Cell wall
- Mitochondria
- Ribosomes
- Nuclear membrane
- Chromatin
- Cellular membrane
- Vacuole
- Cytoplasm
- Chloroplast
- Endoplasmic reticulum

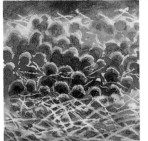

A colony of slime mold

WITHOUT NUCLEUS

Bacteria, previously considered plants, are organisms whose cells lack the nucleus and most organelles. These so-called prokaryotic cells also have cellular walls, although they are not made of cellulose.

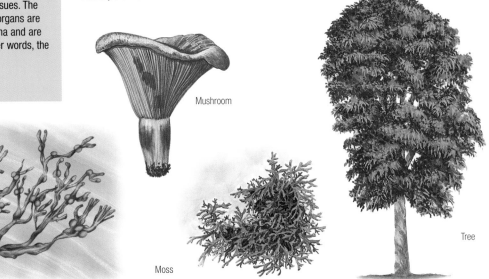

Mushroom

Algae

Moss

Tree

SPECIALIZED CELLS

In one-celled organisms, the **division of one cell** into two cells creates two new individuals. In multicellular organisms, the two new cells remain in association and form part of a growing **tissue**. There are several types of **specialized cells** in a tissue, and the new cells that are formed from them carry out the same functions.

One-celled alga in the process of dividing

GROWTH IN LENGTH OF A STEM AND ROOT

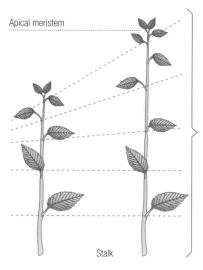

Apical meristem

Area of elongation

Stalk

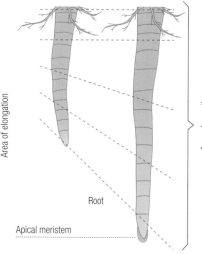

Area of growth

Root

Apical meristem

EMBRYONIC TISSUES

Embryonic tissues are made up of **immature cells** whose main function is to grow, divide, and differentiate, giving rise to other types of tissue. They are located in the growth parts of plants: the points of the roots and stalks contain the **apical meristem**, which produces growth in length; growth in thickness is a function of the **cambium**.

ADULT TISSUES

Adult tissues are made up of **mature cells** that already are specialized in a certain function. There are three types:

1. Protective: These make up the **epidermis** or the external covering of roots, stems, and leaves.

2. Vascular: These are the **xylem** and **phloem**, the conducting vessels of plants. They are made up of a series of microscopic tubes through which water, mineral salts, and nutrients flow.

3. Fundamental: these are the **parenchyma**, **collenchyma**, and **schlerenchyma**, which give sustenance to the plant and help with the production and storage of nutrients. They make up the greatest part of the plant's body.

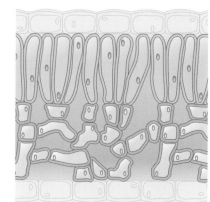

The parenchyma of a leaf

CROSS SECTION OF A PINE TREE TRUNK

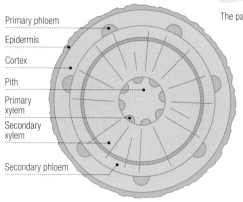

Primary phloem

Epidermis

Cortex

Pith

Primary xylem

Secondary xylem

Secondary phloem

The cells of an earthworm, a human, and an elephant are the same size. It's the **number of cells** instead of their size that causes creatures to be of different sizes.

PLANT ORGANS

The body of complex plants is made up of two basic organic systems: the **roots** and the **shoot** or the aerial part. Both systems are closely connected. The shoot is made up of several organs: the **stem**, the **leaves**, the **flowers**, and the **fruits**.

Flower

Leaves

Fruit

Stem

Root

THE STEM

The stem is the middle part of a plant's body. Algae, fungi, and mosses don't need to have a stem or stalk to hold them up and distribute water and nutrients through conducting vessels. The higher plants, however, need to conduct water, minerals, and nutrients between the leaves and the roots, which they accomplish through the stalk. The other important function of the stalk is to hold the plant's leaves above those of neighboring competitors and keep the plant upright in spite of battering from wind and storms.

THE STRUCTURE OF THE STEM

If you make a transversal cut in a young plant, you will see two areas:

1. The **cortical cylinder**, formed by the **epidermis** and the bark (made up of parenchyma).

2. The **central cylinder**, which has the conducting vessels, the phloem, and the xylem. Filler parenchyma forms the **medulla**, which is the innermost structure of the cylinder.

A STEM IN CROSS SECTION

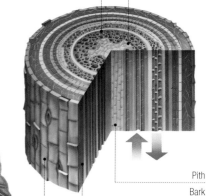

Phloem vessels

Woody vessels (xylem))

Pith

Bark

Epidermis

DIFFERENT TYPES OF STEMS

Type of Stem	Plant
Climbing	Vine
Succulent	Prickly pear
Creeping	Watermelon
Cane	Bamboo
Rhizome	Iris
Bulb	Onion
Tubercle	Potato

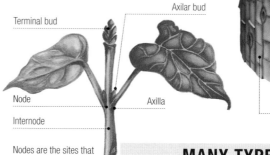

Terminal bud

Axilar bud

Node

Axilla

Internode

Nodes are the sites that give rise to leaves; the internodes are the areas between nodes.

CLIMBING STEMS

Whenever you see a **bean plant** or a **honeysuckle**, imagine how the stem attaches to the pole or other support. These **creeping stems** are called **twiners** or **climbers**. Other stalks climb by means of **clinging roots**, such as **climbing ivy**, of **tendrils**, such as **vines**, or of **thorns**, as with **blackberry bushes.**

MANY TYPES OF STEM

Terrestrial plants can be divided into two major groups based on the consistency of their stems: **herbaceous plants**, which have soft, green stems, and **woody plants**, which have hard, stiff stems, like trees and shrubs. However, based on form and function, a stem may be a **climber**, if it grows up a support; a **succulent**, if it is fleshy and juicy; or a **creeper**, if it grows by supporting itself on the soil. Other stems have common names such as **cane** (a woody stem with nodes), **rhizome** (an underground stem that lives on when the aerial part of the plant dies), **bulb** (a very short stem surrounded by many fleshy leaves), **tuber** (an underground portion of a swollen stem containing nutritional reserves), and others.

Thanks to its climbing stem, ivy scales walls. In the fall, it turns an attractive red.

The stem of an onion is in the shape of a bulb.

Potatoes are swollen underground sections of stalk.

What is commonly called the **leaves** of the **prickly pear** are in fact **succulent stems** that have been converted into organs for storing water. The real leaves have become transformed into spines. As a result, the stems of the prickly pear are green and perform the functions of leaves.

STEM RAMIFICATION

The main stem that sprouts when a plant begins life can branch out by producing secondary stems, which are called **branches**, at the axillas of the leaves. These can branch further, and so on, to produce the phenomenon known as **ramification**. In trees such as the evergreens we bring into the house during the Christmas season, the main axis grows upward and sends out lateral branches. This is typical of **monopodial** ramification. But with most trees, such as the **oak** and the **chestnut**, the growth of the axis ceases early and growth occurs in the lateral branches. This is **sympodial** ramification. In the case of trees, all the branches together constitute the **crown**.

The crown of a chestnut tree covers practically its entire axis.

A sequoia can live more than 3,000 years.

Monopodial ramification allows for increased upward growth. This is the case of the tallest tree in the world, the **giant Sequoia**. Its main axis can reach over 300 feet (100 m) in height.

THE SPONGE TRUNK

The trunk of the **baobab** is an example of a stem adapted to severe droughts, because this is a tree that grows in the hot, arid regions of Africa, India, and Australia. It absorbs and stores great quantities of water that will be of vital importance during the long dry periods. The baobab scarcely exceeds 30 feet (10 m) in height, but it can reach 60 feet (20 m) in circumference.

Thanks to the bark that covers the trunk and the old branches of the cork tree, this tree is resistant to fire, and it sends out new buds that have remained protected from the fire.

The baobab with its characteristic spongy trunk.

PROTECTIVE CORK

Many stems of woody plants lose their primitive green color when they get old and shed their epidermis, which is replaced by a coating of **smooth bark** or, as with the cork tree, **layers of thick, rough cork**. The new tissue, which is made up of **subereous cells** joined together with no spaces between, keeps water from evaporating from the plant. The thick layers also inhibit attacks by parasites, and because of their insulating effect against heat, they protect the plant from very high temperatures.

LEAVES

Why do you suppose leaves are thin and flat? The explanation is that they have to manufacture the plant's food, which they do through **photosynthesis** (synthesis using light). The shape of leaves allows them to absorb the greatest amount of light energy. In addition, you will see that they are arranged on the stem or the branches in such a way that it's as easy as possible for them to capture light.

THE PARTS OF A LEAF

Most leaves consist of three parts: the leaf sheath, the petiole, and the limb. The **leaf sheath** is the insertion base of the leaf into the stem. The **petiole** is the tail of the leaf, which connects the leaf sheath to the **limb**. The latter is the laminar part of the leaf, and it has two faces: the upper, which is called the **face**, and the lower, or **underside**. The petiole continues as the **central vein** of the leaf; it divides into many smaller **veins** that either branch or run parallel to one another.

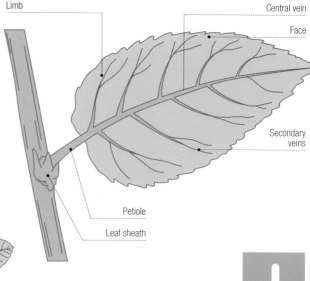

THE PARTS OF A LEAF

Limb

Central vein

Face

Secondary veins

Petiole

Leaf sheath

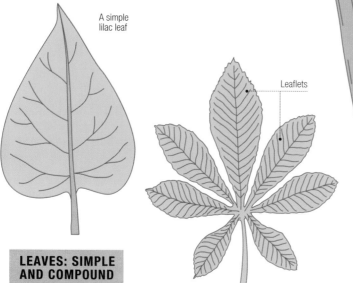

A simple lilac leaf

Leaflets

A palmated (compound) leaf from a horse chestnut tree

LEAVES: SIMPLE AND COMPOUND

A leaf is said to be **simple** when it has a limb in a single piece. When it is made up of various small leaves (**leaflets**) with their small stems branching from a central point or vein, the leaf is said to be **compound**.

In some **tropical palms**, the layer of **wax** that waterproofs the epidermis of the leaves is so thick that it is harvested for use as a polish for shoes and floors.

THE STRUCTURE OF A LEAF

The limb of the leaf is made up of a layer of photosynthetic tissue known as the **mesophyll**, which is covered on both faces by a smooth, lustrous tissue that makes up the leaf's epidermis. The **epidermis** keeps the mesophyll from drying out, so it often is made waterproof by a layer of **wax**. The veins are the **vascular bundles**. The cells of the mesophyll are green because of the large quantity of **chlorophyll** contained in their **chloroplasts**.

The leaves of some plants have small hairs that can fulfill various functions, such as slowing down air currents to reduce evaporation, repelling herbivores (especially if the hairs sting), or reflecting sunlight to prevent overheating of the limb.

CROSS SECTION OF A LEAF

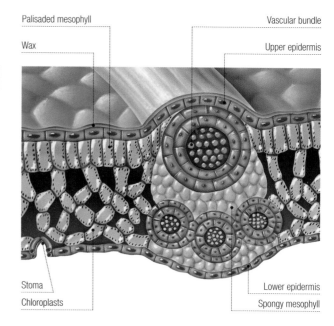

Palisaded mesophyll

Wax

Vascular bundle

Upper epidermis

Stoma

Chloroplasts

Lower epidermis

Spongy mesophyll

THE ESSENTIAL STOMATA

The **stomata** are **microscopic orifices** in the epidermis of a leaf's underside. Their function is to allow entry of **carbon dioxide** from the air, without which photosynthesis could not take place. However, with open stomata, the leaf is at risk of losing water through evaporation. This doesn't happen because each stoma is like a tiny mouth surrounded by a pair of **guard cells** shaped like lips that allow the plant to close its stomata when there is a danger of dehydration.

During the hottest times of the summer days, the leaves of the **eucalyptus** orient themselves parallel to the rays of the sun; that way they heat up less and lose less water through evaporation. It would not be a good idea to plant a eucalyptus as a shade tree for the summer.

A NEARLY CLOSED STOMA (LEFT) AND AN OPEN ONE (RIGHT)

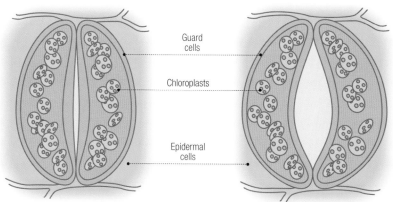

Guard cells

Chloroplasts

Epidermal cells

LEAVES IN DISGUISE

The **scales** of leaf buds and bulbs, many **thorns** or **spines**, and most **parts of a flower** are transformed leaves. That's also true for the **traps of carnivorous plants**, which eat small creatures to get the mineral salts that are not present in the soils where they live.

TYPES OF LEAVES

The shapes that leaves can take are so varied that a list including them all would be almost endless. Botanists distinguish them mainly in these ways:

• **by the limb**: it may **cuneiform** (in the shape of a wedge), **sagittal** (shaped like an arrow), **glabrous** (with a smooth surface), **pubescent** (covered with hairs), etc.

• **by the edge of the limb**: it may be **whole** (smooth edge), **toothed** (with projections), **serrated** (with sharp, angled teeth), **lobed** (divided into rounded sections), etc.

• **by the veins**: it may be **pinnated** (like the fins of a feather), **palmated** (with veins branching from a single point), or **parallel-veined**, among others.

• **by the insertion into the stem**: it may be **sessile** (with no petiole), **sheathed** (the leaf sheath encircles the stem), etc.

The leaves of the main shoots of the **barberry** turn into **spines** that are normally trifurcated.

DIFFERENT TYPES OF LEAVES

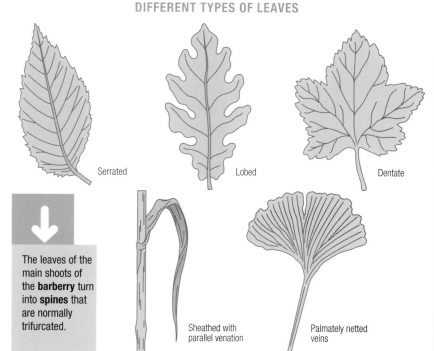

Serrated

Lobed

Dentate

Sheathed with parallel venation

Palmately netted veins

Carnivorous plants in action.

VORACIOUS NEPENTHES

Nepenthes are **carnivorous plants** that attract insects with their bright colors and their nectar. In the vertex of the leaves, there is an excrescence that forms a type of **urn** with a digestive liquid in which the insects drown.

ROOTS

You mustn't think that roots are used only to hold the plant firmly to the ground. All the substances that the plant needs, except for oxygen, carbon dioxide, and solar energy, must be absorbed through the roots that grow in the soil and spread out in search of water and minerals. In addition to absorbing substances from the soil, the roots have to transport them to the stem through the area where these two join, the **root neck**.

THE FEATURES OF A ROOT

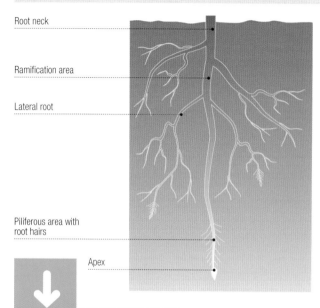

Root neck

Ramification area

Lateral root

Piliferous area with root hairs

Apex

In general, roots have two well-defined, important areas: the apex or vegetative cone and the piliferous area. The **apex** is the growth area located at the end of the root. It is very short (around a fifth of an inch [5 mm]); if it weren't, the root would easily become twisted as it grew in opposition to the resistance of the soil. In addition, the root is protected by a type of hood known as the **calyptra** or root cap, which helps it penetrate the soil. The piliferous area is the youngest part of the root; it begins a few millimeters from the calyptra and is covered with **root hairs**.

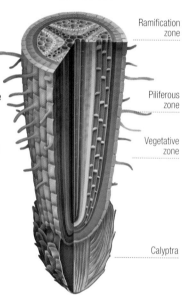

ENLARGEMENT OF A ROOT APEX

Ramification zone

Piliferous zone

Vegetative zone

Calyptra

Except in special areas such as deserts and in water, the total mass of a plant's roots is approximately equal to that of its branches.

The roots of all plants in the world extract several tons (thousands of pounds [kilos]) of minerals from the soil every minute.

DETAIL OF THE PILIFEROUS ZONE

ROOT HAIRS

The root hairs increase the root's absorption surface, thereby facilitating the absorption of water and minerals from the soil. However, they live for only a few days, because they cover only a very small part of the root. The oldest ones dry up and fall off and are replaced by new ones that form near the apex. These also are called **absorbent hairs** because water and dissolved substances enter the roots through their fine membrane and are conducted to the vascular bundles that deliver them to the plant's leaves.

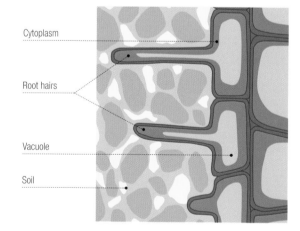

Cytoplasm

Root hairs

Vacuole

Soil

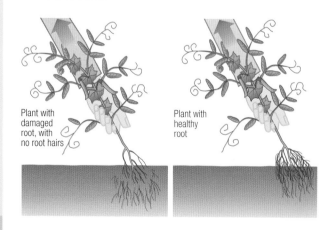

Plant with damaged root, with no root hairs

Plant with healthy root

TRANSPLANTATION DANGERS

If you are not careful in transplanting and the root hairs or the rhizodermis of the root is lost, thereby leaving it exposed, it will be very difficult for the plant to continue to develop and even survive, because its roots will be deprived, for several days, of their ability to absorb nutrients. The only way to save the plant is to remove its leaves and reduce dehydration until new root hairs can grow.

Introduction

Plant Anatomy

Plant Physiology

Reproduction

Flower, Fruit, and Seed

Ecology and Evolution

Algae

Fungi

Plants

Plants with Flowers and Fruits

Plants and Their Environment

Aquatic Plants

Wild Plants

Domesticated Plants

Gardens

Alphabetical Subject Index

THE STRUCTURE OF THE ROOT

Roots are mostly composed of **bark**, and especially of **parenchymal tissue**. The thickest roots greatly resemble the branches of woody plants, but the smallest ones are covered only by a **radicular epidermis** or **rhizodermis**, from the cells of which the **root hairs** extend. The interior of the root is taken up by the **central** cylinder, through which run the vascular bundles of xylem and phloem.

THE PARTS OF A ROOT

Xylem

Phloem

Central cylinder

Parenchyma

Bark

Root hairs

Rhizodermis

The pagoda ivy sends out **aerial roots** that attach to the soil and act as real "stilts" that hold up the large horizontal branches.

Just as there are tubers that are parts of stems, such as potatoes, there are also **root tubers**, such as the tiger nuts that are used in Spain to make the drink horchata.

TYPES OF ROOTS

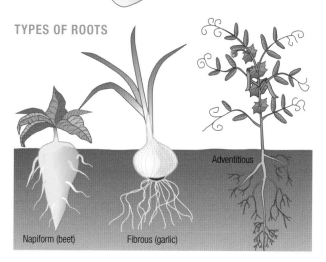

Napiform (beet)   Fibrous (garlic)   Adventitious

ABNORMAL BUT VERY USEFUL ROOTS

When a seed germinates, the root and the shoot develop in tandem. But roots also can originate advantageously on the stem or the leaves of an adult plant. These roots, which are known as adventitious roots, can be part of a plant's development, as in the case of the stolons of the strawberry plant or the hooks of the ivy, and they also can be started artificially for the purpose of propagating plants through cuttings.

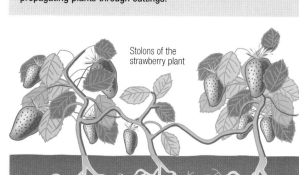

Stolons of the strawberry plant

TYPES OF ROOTS

When secondary roots branch off a main root that is developing underground, that is an **adventitious root**. In a root system, if there is no distinction between the main root and the secondary ones, the roots are said to be **fibrous**. There also are specialized roots that store food substances, such as the **napiform roots** of the beet, the carrot, and the turnip, as well as **aerial roots** that have different purposes (support, respiration, etc.).

Mangroves are plants typical to littoral regions of tropical areas known as mangrove swamps. Mangrove roots, which spread out in the mud, send out ascending ramifications that project out of the water as if they were snorkels. These are respiratory roots.

PHOTOSYNTHESIS

In nature there are two basically different types of beings: those that make their own food and those that live off other organisms, whether living or dead. The first category is the plants, algae, and certain bacteria; the second is the fungi, the majority of bacteria, and all the animals, including the protozoa. The organisms that feed by themselves carry out photosynthesis, in other words, they synthesize organic material on the basis of mineral matter by using the energy in the sun's light.

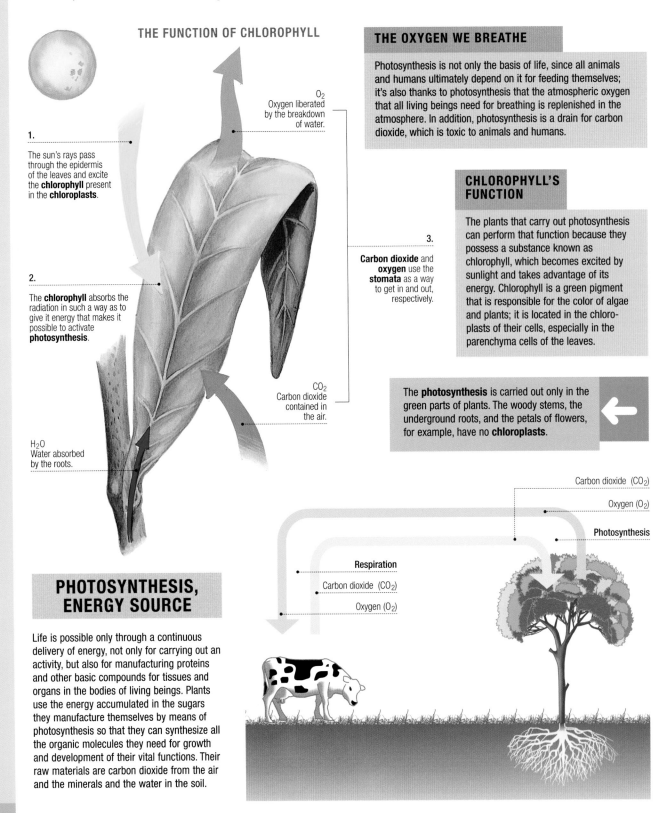

THE FUNCTION OF CHLOROPHYLL

1.

The sun's rays pass through the epidermis of the leaves and excite the **chlorophyll** present in the **chloroplasts**.

2.

The **chlorophyll** absorbs the radiation in such a way as to give it energy that makes it possible to activate **photosynthesis**.

O_2
Oxygen liberated by the breakdown of water.

3.

Carbon dioxide and **oxygen** use the **stomata** as a way to get in and out, respectively.

CO_2
Carbon dioxide contained in the air.

H_2O
Water absorbed by the roots.

THE OXYGEN WE BREATHE

Photosynthesis is not only the basis of life, since all animals and humans ultimately depend on it for feeding themselves; it's also thanks to photosynthesis that the atmospheric oxygen that all living beings need for breathing is replenished in the atmosphere. In addition, photosynthesis is a drain for carbon dioxide, which is toxic to animals and humans.

CHLOROPHYLL'S FUNCTION

The plants that carry out photosynthesis can perform that function because they possess a substance known as chlorophyll, which becomes excited by sunlight and takes advantage of its energy. Chlorophyll is a green pigment that is responsible for the color of algae and plants; it is located in the chloroplasts of their cells, especially in the parenchyma cells of the leaves.

The **photosynthesis** is carried out only in the green parts of plants. The woody stems, the underground roots, and the petals of flowers, for example, have no **chloroplasts**.

Carbon dioxide (CO_2)

Oxygen (O_2)

Photosynthesis

Respiration

Carbon dioxide (CO_2)

Oxygen (O_2)

PHOTOSYNTHESIS, ENERGY SOURCE

Life is possible only through a continuous delivery of energy, not only for carrying out an activity, but also for manufacturing proteins and other basic compounds for tissues and organs in the bodies of living beings. Plants use the energy accumulated in the sugars they manufacture themselves by means of photosynthesis so that they can synthesize all the organic molecules they need for growth and development of their vital functions. Their raw materials are carbon dioxide from the air and the minerals and the water in the soil.

THE BASIC UNIT OF PHOTOSYNTHESIS

The **chloroplast** is filled with tiny vesicles known as **tilacoids**, which are the basic units of photosynthesis because the **chlorophyll** is located inside their membranes. The tilacoids are grouped in **grana**. Each granum contains many tilacoids piled one on top of another.

A PROGRESSIVELY MAGNIFIED VIEW OF CHLOROPLASTS

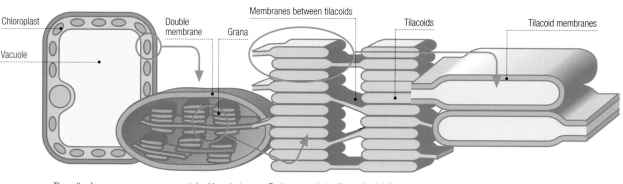

Chloroplast

Vacuole

Double membrane

Grana

Membranes between tilacoids

Tilacoids

Tilacoid membranes

The cells of green plants . . .

. . . contain chloroplasts made up of grana.

Each granum looks like a pile of deflated balls arranged in perfect order; these are the tilacoids.

The tilacoids of neighboring grana are interconnected by membranes.

STORING CHLOROPHYLL

You probably have noticed that in the fall many trees turn yellow or red before they shed their leaves. This is because the chloroplasts contain other **photosynthetic pigments** besides chlorophyll. The plants that lose their leaves in the fall display these pigments before the arrival of winter because they withdraw the chlorophyll from their leaves and store it in their permanent tissues before dropping their leaves, which then contain only the accessory pigments in beautiful reddish hues. In the springtime they dip into the stored chlorophyll.

→ Photosynthesis intensifies in proportion to an increase in the concentration of carbon dioxide in the air, the temperature (up to a certain point), and the intensity of light. Photosynthesis does not occur in the dark.

Grass and trees grow by using carbon dioxide in the air and water, plus the minerals in the soil, with the aid of the sun.

FALL

SPRING

↓ Just like animals, plants also breathe: they absorb atmospheric oxygen and give off carbon dioxide. But their photosynthetic activity is far greater than their respiratory activity.

PLANT NUTRITION

Photosynthesized sugars already contain three basic elements of living matter: carbon, oxygen, and hydrogen. But, in order to form their tissues and organs, plants also need other elements that they must absorb, either directly from water (algae) or through their roots. The fungi practice another type of nutrition; they absorb dissolved organic materials through their membranes.

HOW PLANTS GET THEIR NUTRIENTS

Land plants get their nourishment by absorbing water and dissolved mineral salts (**crude sap**) through the root hairs and pumping it toward the leaves, where all the organic compounds that the plant needs for growth and reproduction are manufactured. The liquid that contains these compounds, plus the ones manufactured through photosynthesis, makes up the **elaborated sap** that gets distributed throughout the plant.

Many of the world's agricultural problems are caused by soils that are deficient in nitrogen.

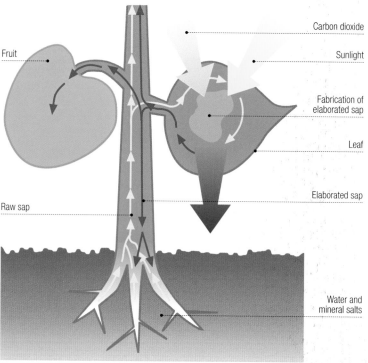

THE NUTRIENT CYCLE OF PLANTS

Fruit

Carbon dioxide

Sunlight

Fabrication of elaborated sap

Leaf

Elaborated sap

Raw sap

Water and mineral salts

THE MECHANISM FOR ABSORPTION AND TRANSPIRATION

H_2O Transpiration (evaporation of water contained in the leaf)

Rising of crude sap

Absorption of the solution from the soil in the piliferous area

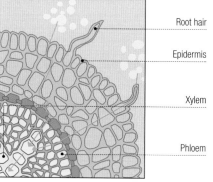

Root hair

Epidermis

Xylem

Phloem

Water loss by transpiration through the leaves produces a suction that causes the raw sap to rise from the xylem of the roots.

When there is a nutrient deficiency, the plant's growth is limited, even though there may be an excess of other nutrients.

CAPTURING AND TRANSPORTING MINERAL SALTS

Mineral salts in the soil can be absorbed by the plant only when they are dissolved in water, since solid particles can't get through the cell membranes of the roots. The solution passes through the epidermis and the bark, moving from one cell to another and through the cell walls without penetrating the cells. That's how it reaches the xylem and starts its ascent through the roots and continues up the stem.

Introduction

Plant
Anatomy

**Plant
Physiology**

Reproduction

Flower, Fruit,
and Seed

Ecology and
Evolution

Algae

Fungi

Plants

Plants with
Flowers and
Fruits

Plants and Their
Environment

Aquatic Plants

Wild Plants

Domesticated
Plants

Gardens

Alphabetical
Subject Index

A PLANT'S NUTRITIONAL NEEDS

Among all the elements that a plant absorbs from the soil, a few are necessary to all plants and in fairly large quantities. That's because they are used in the basic units that make up the tissues, organs, and important substances for the plant, or else because the plant uses a lot of them to function properly. Other nutrients are likewise necessary but in very small quantities.

NUTRIENTS THAT ALL PLANTS NEED (WITH THEIR CHEMICAL SYMBOLS)

Elements required in large amounts and present in all tissues and organs	Carbon (C) (from atmospheric carbon dioxide)	Basic component of all organic molecules
	Oxygen (O)	Associated with C or H in organic molecules
	Hydrogen (H)	Associated with C or O in organic molecules
	Nitrogen (N)	Basic component of proteins in all living beings
Elements required in large amounts	Potassium (K)	Necessary for proper plant functioning
	Phosphorus (P)	Necessary for proper plant functioning and an indispensable component of chromosomes
	Sulfur (S)	Essential component of proteins
	Calcium (Ca)	Important component of cell walls
	Magnesium (Mg)	Essential component of chlorophyll
Elements required in very small amounts	Iron (Fe)	Essential component of chlorophyll
	Boron (B)	Necessary for plant functioning
	Zinc (Zn)	Necessary for plant functioning
	Manganese (Mn)	Necessary for plant functioning
	Chlorine (Cl)	Necessary for plant functioning
	Molybdenum (Mo)	Necessary for plant functioning
	Copper (Cu)	Necessary for plant functioning

Fertilizers improve the physical, chemical, and biological properties of arable soil, promoting plant nutrition.

WHAT ARE FERTILIZERS?

A fertilizer is a nutrient or a mixture of nutrients that is applied to the soil to make up for some deficiency. The fertilizer must be inorganic so it can be used by the plant. But it also can be incorporated in organic form (humus or manure) so that the plant can use it as it decomposes and mineralizes.

Earthworms eat plant remains and break them down in their digestive tube.

THE NITROGEN CYCLE

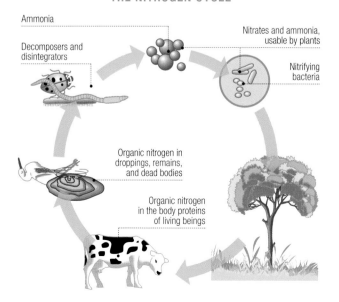

Ammonia

Decomposers and disintegrators

Nitrates and ammonia, usable by plants

Nitrifying bacteria

Organic nitrogen in droppings, remains, and dead bodies

Organic nitrogen in the body proteins of living beings

THE MINERALIZATION OF ORGANIC MATERIAL IN THE SOIL

The organic material in the soil, the **humus**, is made up of the remains of partially decayed plant and animal remains. This humus is especially rich in elements that plants need, such as sulfur and calcium, but especially in nitrogen. However, this must first be converted to mineral nitrogen in the form of **nitrates** or **ammonia** so that the plants can use it.

ALLIES OF THE PLANTS

Nitrogen in a form that plants can assimilate is a rare commodity in most soils. But plants have some allies that convert the organic nitrogen into nitrates and ammonia. These are the consumers of organic remains (**detritophagous feeders**), such as earthworms and millipedes, and a multitude of microorganisms that break down organic matter (fungi and bacteria) and **nitrifying bacteria**.

GROWTH AND DEVELOPMENT

In the development process of every living being, the basic phenomenon is the growth of each cell. But this growth has a limit. Multicelled plants develop from a fertilized female reproductive cell known as a **zygote**, which keeps dividing. The majority of the newly created cells expand and differentiate to form tissues and organs that make the individual grow; however, there are always a few cells in locations known as **vegetative points** that never lose their capacity to form new parts of the plant.

THE GROWTH OF ALGAE

Multicelled algae increase in size through two basic types of growth: **generalized growth**, where all cells are capable of dividing, and **localized growth**, in which cellular division is restricted to certain parts of the alga. The latter case may involve **apical growth** or **intercalary growth**.

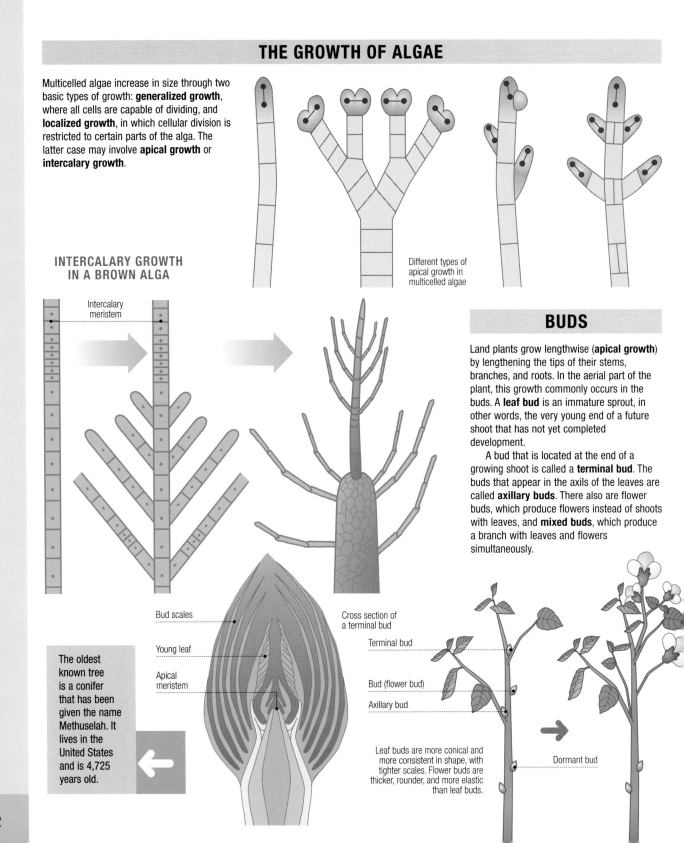

Different types of apical growth in multicelled algae

INTERCALARY GROWTH IN A BROWN ALGA

Intercalary meristem

BUDS

Land plants grow lengthwise (**apical growth**) by lengthening the tips of their stems, branches, and roots. In the aerial part of the plant, this growth commonly occurs in the buds. A **leaf bud** is an immature sprout, in other words, the very young end of a future shoot that has not yet completed development.

A bud that is located at the end of a growing shoot is called a **terminal bud**. The buds that appear in the axils of the leaves are called **axillary buds**. There also are flower buds, which produce flowers instead of shoots with leaves, and **mixed buds**, which produce a branch with leaves and flowers simultaneously.

Bud scales

Young leaf

Apical meristem

Cross section of a terminal bud

Terminal bud

Bud (flower bud)

Axillary bud

The oldest known tree is a conifer that has been given the name Methuselah. It lives in the United States and is 4,725 years old.

Leaf buds are more conical and more consistent in shape, with tighter scales. Flower buds are thicker, rounder, and more elastic than leaf buds.

Dormant bud

APICAL DOMINANCE

As long as the **apical meristem** of the stem is intact, the lateral buds commonly remain fairly "dormant," and, as a result, **lateral ramification** tends to be suppressed in favor of **apical growth**. This phenomenon, which is called apical dominance, is the reason why many plants fill out with branches when their tips are pruned.

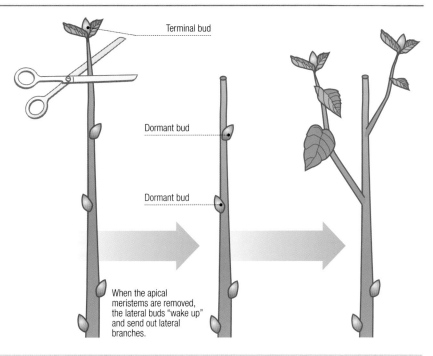

Terminal bud

Dormant bud

Dormant bud

When the apical meristems are removed, the lateral buds "wake up" and send out lateral branches.

Developing all buds would be a waste for a plant. The **dormant buds** at the base of the branches sprout only if the branch breaks or dies of old age.

GROWTH RINGS

More complex plants also increase in thickness. In regions that have clearly defined seasons, this makes it possible to determine the age of a tree by examining a cross section of the trunk. **Annual growth** is embodied in the rings formed by the successive layers of wood. Each year a layer of wood (xylem) forms from the **cambium**, and another layer, of bast (phloem), forms toward the outside. The **wood rings** are distinguishable because the conducting vessels formed in the summer are smaller and tighter, so they form a narrow, dark circle; the ones that form in the spring, on the other hand, are lighter in color.

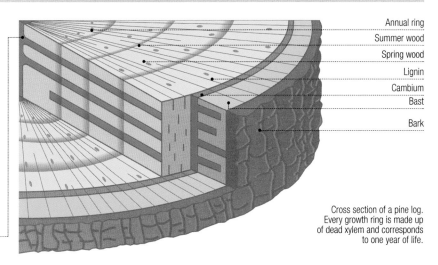

Annual ring
Summer wood
Spring wood
Lignin
Cambium
Bast
Bark

Pith

Cross section of a pine log. Every growth ring is made up of dead xylem and corresponds to one year of life.

HOW MANY YEARS DOES A PLANT LIVE?

The length of a plant's life varies according to species. **Annual plants** sprout, grow, flower, produce seeds, and die in less than a year. **Biennial plants** live for two years. **Perennial plants** live for several years (and some of them for many years) and may bloom every year.

The part of a perennial plant that ages most clearly is the leaves. Even with evergreen plants, there is a constant shedding of leaves.

LIFE EXPECTANCY OF SOME PERENNIAL PLANTS

Plant	Possible Life Span (in years)
Blueberry	28
Poplar	500
Elm	600
Lime	1,000
Oak	1,000
Yew	3,000
Giant sequoia	5,000

Onions are a biennial plant: they accumulate reserves in their bulb during the first year in anticipation of flowering the following year, and then they die.

The blueberry is a perennial plant.

The poppy is an annual plant.

REPRODUCTION AND HEREDITY

The main feature of **reproduction** is the formation of a new generation of descendents that resemble the members of the preceding generation. This requires transmission of the traits and characteristics of the parents to the offspring, through the **genes**, which is known as **heredity**. The science that studies the structure, transmission, and expression of genes is **genetics**.

CHROMOSOMES AND GENES

Chromosomes are long filaments of **DNA** (deoxyribonucleic acid) and other proteins. Genes are small segments of DNA that codify a type of **biological information** and together cause individuals to resemble their progenitors. Every individual of every species contains a particular number of chromosomes in every cell of its body. But chromosomes normally occur in pairs, so there are two chromosomes of every type (**homologous chromosomes**) that carry information that corresponds to the same characteristics, although not necessarily the same information.

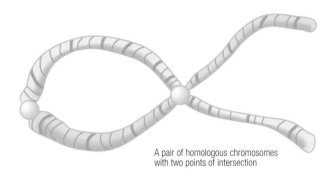

A pair of homologous chromosomes with two points of intersection

Every characteristic of a plant, such as the bean, is determined by one gene, whether the shape of its leaves, the color of the flower, the type of its fruit, or the minimum temperature needed by the seeds for germination.

 Every chromosome is composed of DNA, where the genes are located. Every chromosome can contain hundreds or even thousands of genes.

DIFFERENT WAYS OF REPRODUCING

In order to produce new individuals, plants can use two types of reproduction:

• **Asexual:** Cells, groups of cells, or fragments with germinating ability detach from the mother plant, germinate directly, and give rise to new independent beings; or

• **Sexual:** Two classes, male and female, of specialized germinating cells are produced; these are called **gametes** and may occur in a single individual or in separate ones. The new being arises through fusion of a male gamete (antherozoid) with a female one (ovule), that is, through fertilization.

THE PHASES OF SEXUAL REPRODUCTION

Production of gametes

   Fusion of gametes

Cell of progenitor with two sets of chromosomes (2n chromosomes)

Division with redistribution and crossing of chromosomes

Formation of four gametes with a single set of chromosomes (n) in each

Male gamete (n) × female gamete (n) → zygote (2n) with genes from father and mother

GENETIC ENGINEERING

Genetic engineering is a series of laboratory techniques that make it possible to manipulate the DNA or the genes of a living being, including introducing new genes into its chromosomes and producing **transgenics**. Corn and rice are the plants that have been used to produce the greatest number of transgenics. One of them, golden corn, has had genes introduced that increase its vitamin A content.

An ear of corn

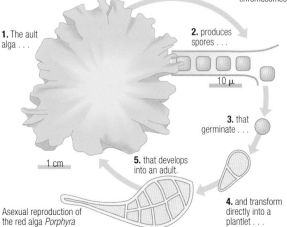

1. The ault alga . . .

2. produces spores . . .

10 μ

3. that germinate . . .

4. and transform directly into a plantlet . . .

5. that develops into an adult.

1 cm

Asexual reproduction of the red alga *Porphyra*

Introduction

Plant Anatomy

Plant Physiology

Reproduction

Flower, Fruit, and Seed

Ecology and Evolution

Algae

Fungi

Plants

Plants with Flowers and Fruits

Plants and Their Environment

Aquatic Plants

Wild Plants

Domesticated Plants

Gardens

Alphabetical Subject Index

USING BOTH TYPES OF REPRODUCTION

Most plants use both types of reproduction in such a way that a generation of asexual reproduction using **spores**, known as **sporophytic**, alternates with a sexual generation that reproduces through **gametes** and is referred to as **gametophytic**. Both generations commonly are represented by individuals that have a completely different appearance but are not always independent of one another. There are many modalities of **alternation of generations**.

In the life cycle of seed-producing plants, the sporophyte, the plant that we see, is entirely dominant. The gametophytes are very small and remain invisible inside the flower: the producer of male gametes, inside a **grain of pollen**; and the female, inside the **embryonic sac**. After fusion of the gametes, a **seed** is formed, which is the origin of the sporophyte.

ALTERNATION OF GENERATIONS IN PLANTS THAT PRODUCE SEEDS

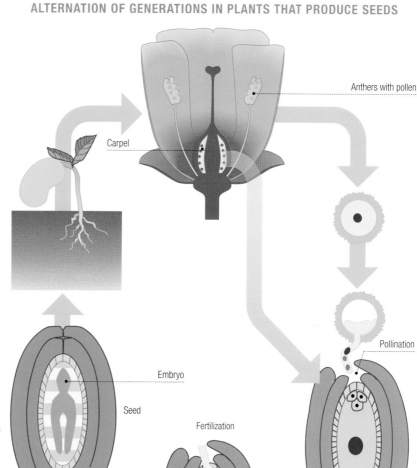

Anthers with pollen

Carpel

Pollination

Fertilization

Embryo

Seed

MENDEL'S EXPERIMENT
(T = tall gene; t = dwarf gene)

TT

tt

First-generation hybrids

Tt

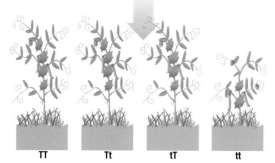

TT   Tt   tT   tt

The tallness gene was dominant over the dwarfism gene.

The individuals in any generation exhibit differences from one another and from their progenitors, which are known as **variations**. Variations caused by environmental factors in the development of the organism are not hereditary. Only variations of genetic origin are passed on.

HEREDITY AND ITS LAWS

The laws that govern the transmission of hereditary characteristics were discovered by Gregor Mendel (1822–1884). In one of his experiments, he crossed a tall pea plant with a short one, and all the offspring grew tall. But when these were crossed with one another, the next generation included tall and dwarf plants in a proportion of three to one.

ASEXUAL REPRODUCTION

In nature, plants reproduce asexually in such a way that a single progenitor divides, forms buds, or fragments, to give rise to two or more descendents, or else produces spores that germinate directly.

In all cases, this is a quick process, which allows individuals well adapted to their environment to produce new generations of individuals equally well adapted, because they possess identical genes.

THE SIMPLEST AND QUICKEST MULTIPLICATION

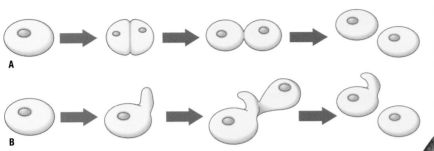

A

B

Many one-celled algae reproduce by simple **division**, that is, by dividing their cell into two parts (**bipartition**) and separating from the resulting offspring. Yeasts adopt a variant of this system that's known as **gemmation**: the cell forms a type of yeast similar to a wart, which grows until it detaches from the mother and becomes independent.

Multiplication by bipartition (A) and gemmation (B)

In a single day, a one-celled organism that reproduces by **bipartition** or **gemmation** can produce several million descendents.

WHY IS BREAD SPONGY?

If you look at the bread we eat, you will see that one of its good qualities is its sponginess. This characteristic is caused by the **yeast** that is added to the dough several hours before it's put into the oven. This yeast is a cultivation of **one-celled fungi** that reproduce by **gemmation** at astonishing speeds. They cause the dough to **ferment**, give off carbon dioxide, increase the volume of the dough, and produce a spongy texture.

Bread has a spongy texture thanks to yeast.

FRAGMENTATION

Many multicellled plants multiply without developing specialized reproductive organs. Spontaneously, or as the result of some external mechanical influence, parts of its body separate and give rise to new individuals.

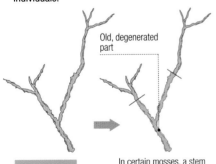

Old, degenerated part

In certain mosses, a stem ramifies, and when its oldest parts degenerate, several branches become independent.

The brewing of **beer** is another example of rapid **multiplication by gemmation**. A **yeast** similar to the ones used in breads is added to **germinated barley** to cause the fermentation that produces beer.

BASKETS OF PROPAGULES

Some lower plants, such as the liverworts, produce little basket-like structures in the top part of the stem that contain groups of germinating cells known as **propagules**.

When it rains, the drops of water that fall on these little baskets dislodge and carry away these propagules, which turn into new plants.

A BASKET OF PROPAGULES

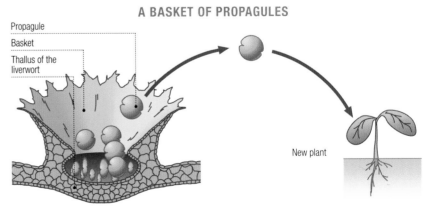

Propagule
Basket
Thallus of the liverwort

New plant

Introduction

Plant
Anatomy

Plant
Physiology

Reproduction

Flower, Fruit,
and Seed

Ecology and
Evolution

Algae

Fungi

Plants

Plants with
Flowers and
Fruits

Plants and Their
Environment

Aquatic Plants

Wild Plants

Domesticated
Plants

Gardens

Alphabetical
Subject Index

PLANTS THAT GIVE OFF OFFSPRING

Certain plants produce special **buds**, either on their stalks or on their leaves, that turn into little plantlets that are already provided with an incipient root. These buds, which are known as **bulbils**, begin to develop right on the mother plant, and, when they fall to earth, they are capable of living as independent plants. There also are some plants that accumulate bulbils on their roots.

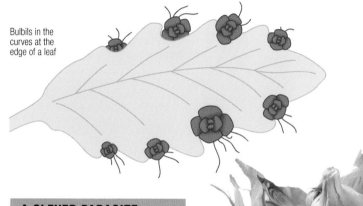

Bulbils in the curves at the edge of a leaf

Axillary bulbils

A CLEVER PARASITE

The fungus that parasitizes vines and causes the disease known as **mildew** reproduces by **mobile spores** that are adapted to aquatic life. However, the mildew develops in a dry environment. How can that be? Its spores take advantage of the morning dew to germinate and develop.

The heads of garlic that we use in cooking are made up of bulbils, which are what we refer to as the garlic cloves. Garlic is cultivated by planting these cloves.

Axillary bulbils of *Dentaria*

SPORES

Different types of mobile spores or planospores. There are planospores with one or several flagella, smooth or barbed, oriented toward the front, or opposed.

The most common form of asexual reproduction among plants is **sporulation**, which involves the production of special cells known as **spores** by successive divisions inside a mother cell known as a **sporangium**. There are **mobile spores** or **planospores** and **immobile spores** or **aplanospores**. The former are characteristic of aquatic plants, and the latter are for multiplication outside the liquid medium.

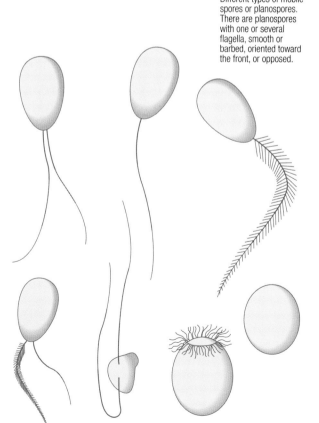

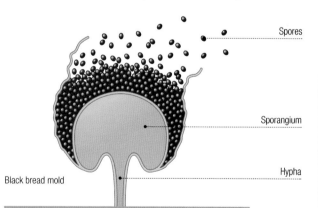

Spores

Sporangium

Hypha

Black bread mold

WHAT WE SEE ARE THE SPORANGIA

In all likelihood you have often seen fruits, bread, and other foods covered with a white, green, or black fuzz. That fuzz is the **sporangia** of the fungus that has invaded the food with its filaments or **hyphae**. These sporangia are filled with immobile **spores** that are dispersed by the wind when the hyphae open. The black color of bread mold, for example, is caused by the color of the spores contained in the sporangia, which is the part that we see.

ARTIFICIAL ASEXUAL REPRODUCTION

Both farmers and scientific researchers frequently use the regenerative capacity of plants in making them multiply. That's how they manage to propagate strains that can't reproduce sexually, as well as produce descendents with characteristics identical to those of a specific individual, which would not happen through sexual reproduction.

FIRST ROOT, THEN WEAN

Farmers have always used the ability of aerial shoots to take root to obtain plants through **layering**. This can be done in many ways, but in essence it involves burying part of a young branch, bending it if necessary, and forcing it to put out roots without separating it from the mother plant. Once the roots are out, the shoot can feed itself, and it is "weaned," that is, it is separated from the mother plant.

DIFFERENT TYPES OF LAYERING

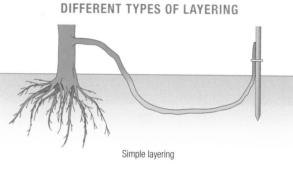

Simple layering

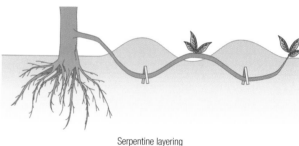

Serpentine layering

Layering by hilling

The aerial layering is covered with humus, which is tied on to keep it in place, and it is wrapped in plastic to hold in the moisture.

Multiplication through shoots. Plants that send out shoots at their base also can be propagated by removing the shoots with a little bit of root and planting them.

A small greenhouse is a basic element in success with cuttings, since rooting requires warmth and moisture.

Chinese layering

PROPAGATION THROUGH CUTTINGS

Cutting involves reproducing a plant from a piece of stem, leaf, or root taken from the plant. It is a very commonly used technique because it allows the production of a great number of offspring very simply and cheaply. The piece of stem must contain buds, and the part that's buried is sprayed with plant growth hormones to encourage rooting.

Cutting in water

Begonias are easy to multiply by using the leaf as a cutting.

MYSTERIOUS COCONUT MILK AND HORMONES

The tasty milk that's contained in ripe coconuts was used in cultivating plant cells, although the causes for its positive effects were not known. Later it was discovered that the milk contains one of the **growth hormones** of plants. Nowadays you can buy these hormones in gardening supply stores. Apply them to the ends of the cuttings that are buried and you will almost surely experience success.

Introduction

Plant
Anatomy

Plant
Physiology

Reproduction

Flower, Fruit,
and Seed

Ecology and
Evolution

Algae

Fungi

Plants

Plants with
Flowers and
Fruits

Plants and Their
Environment

Aquatic Plants

Wild Plants

Domesticated
Plants

Gardens

Alphabetical
Subject Index

GRAFTING

Grafting involves inserting a piece of another plant containing buds into a rooted plant so that the tissues of both plants **fuse** together. The plant receiving the graft is known as the **rootstock**, and the inserted part is the **scion** or **graft**. If the fusing is successful, the **conductive** vessels of the scion communicate with those of the rootstock so that the scion gets the nutrients it needs for growth.

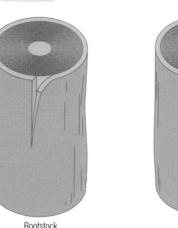
Scion

Rootstock

Crown graft

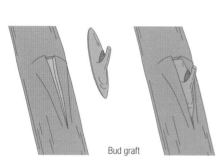
Bud graft

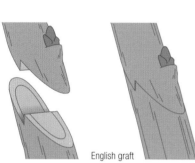
English graft

Types of scions. Whatever technique is used, the goal is always the same: to produce a specimen that uses the roots of the stock and an aerial part identical to the plant from which the scion was taken.

In the past, all oranges had seeds.

THE MOTHER OF ALL SEEDLESS ORANGES

All seedless oranges in the world come from trees that derive from an original orange tree that arose spontaneously in a Brazilian orchard in the nineteenth century. The originality of that orange tree was caused by a genetic alteration beyond the orchardist's control. The orange tree was grafted to other varieties of citrus trees, and that's how we now have seedless oranges.

↓
If you want to practice grafting, you need to keep in mind that only plants with very similar structures can be fused together, such as an orange and a lemon tree, or an almond and a peach.

ADVANTAGES OF PRODUCING CLONED INDIVIDUALS

If you had a patch of **strawberries** and noticed that one plant grew healthier and produced better fruit than the others, you would want to produce **cloned descendents** from that plant; in other words, strawberry plants that are genetically identical to your favorite one. This is what the good strawberry cultivators do who want to ensure the quality of their production: even though the planting is more costly, they buy strawberry plants produced by **micropropagation**.

REPRODUCTION IN A LABORATORY

The most modern technique of propagating consists of **laboratory cultivation** of **meristems** and **bud apices**, which is known as **micropropagation** or tissue culture. It is an advantageous cultivation method, and, with adequate quantities of plant hormones, roots and shoots quickly differentiate from the initial undifferentiated tissue. This is a very quick process that takes up little room and can be used to easily produce thousands of **cloned individuals** at very low prices and, especially, free of viruses and fungi that cause plant diseases.

Laboratory work makes it possible to produce specimens with identical characteristics at very low prices and with resistance to diseases. The photograph shows an experimental orchid greenhouse in Thailand.

FLOWERS

A flower is a set of specialized structures used in **sexual reproduction**. In most flowers there are four parts known as **verticils**: the **calyx**, the **corolla**, the **androecium**, and the **gynecium**. The latter two are the reproductive organs; the function of the calyx and the corolla is to protect these organs and attract the insects that can aid in reproduction.

THE CALYX AND THE COROLLA

Usually the most attractive part of a flower is the **corolla**, which is made up of leaves that have become transformed into brightly colored **petals**. The corolla is surrounded by smaller leaves known as **sepals**, which together make up the **calyx,** and, in some flowers, are joined together to form a single piece. Before a flower develops, the **flower bud** is covered and protected by the calyx.

COLOR STRATEGY AND SUGAR

If we look at the interior of a **violet** with a magnifying glass, we see the **nectaries** that contain the sugary **nectar** that bees and other **pollinating insects** love so much. But how do these insects find something that's so well hidden? The flower takes it upon itself to help them learn where the nectar is located by guiding them toward the nectarines with bright color contrasts. Recall, too, that bees see colors that we don't.

Different types of stamens

THE MALE ORGANS

The flower's male reproductive organs are the **stamens**, which are located inside the corolla and together make up the **androecium**. Each stamen consists of a **filament** that ends in an **anther** made up of two lobules with two **pollen sacs** inside; they contain the **grains of pollen** that produce the yellowish color that we see in the center of flowers. Their function is to produce **male gametes**. When they reach maturity, the pollen sacs break open and release the pollen.

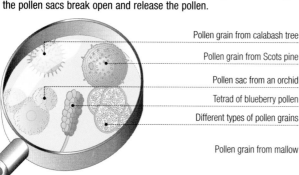

Pollen grain from calabash tree

Pollen grain from Scots pine

Pollen sac from an orchid

Tetrad of blueberry pollen

Different types of pollen grains

Pollen grain from mallow

THE PARTS OF A TYPICAL FLOWER

Anther

Stigma

Stamens

Petals

Style

Sepals

Ovary

Receptacle

The most expensive spice in the world is the stigmas of the saffron flower. They must be harvested by hand.

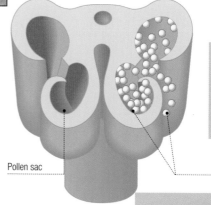

Pollen sac

Pollen grains

Cross section of an anther

The science that studies pollen grains is called **palynology**.

POLLEN AND POLICE INVESTIGATIONS

Sherlock Holmes knew that no two species of plants have the same pollen grains. So the pollen found on the clothing of suspects can be a valid trail to the location of the crime. This characteristic of pollen grains is very useful for prehistoric environments, too, since the hard coating on the pollen grains resists fossilization.

Introduction

Plant
Anatomy

Plant
Physiology

Reproduction

**Flower, Fruit,
and Seed**

Ecology and
Evolution

Algae

Fungi

Plants

Plants with
Flowers and
Fruits

Plants and Their
Environment

Aquatic Plants

Wild Plants

Domesticated
Plants

Gardens

Alphabetical
Subject Index

THE FEMALE REPRODUCTIVE ORGAN

The innermost verticil of the flower is the **gynecium**, which consists of one or several folded leaves united at their borders, known as **pistils**. The gynecium almost always consists of several carpels joined in a single piece known as a pistil because of its resemblance to a mortar (*pistillum* in Latin). The swollen part below is the ovary, which contains the **ovules**; the "sleeve" is the **style**, and the "head of the sleeve" is the **stigma**. The stigma commonly produces a sugary and sticky liquid to which the pollen grains adhere.

Three different arrangements of ovules in the ovary. The role of the **stigma** is to receive the pollen. The **style** conducts the pollen to the ovary. Fertilization takes place in the **ovary**.

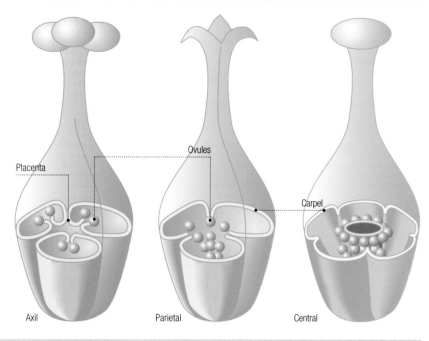

Placenta

Ovules

Carpel

Axil   Parietal   Central

THE SEX OF THE FLOWER AND THE SEX OF THE PLANT

Most flowers are **hermaphrodites**, in other words, they have both stamens and pistils. But there are some species that are **monoeicious**, in which the same plant has two types of flowers: some male ones (without pistils) and others that are female (without stamens). There are also **dioecious** plants, that is, with male individuals (with only male flowers) and female individuals (with only female flowers).

→ Plants of the arum species change sex with age. They are male when they are young, and, when they grow old, they have only flowers with pistils.

INFLORESCENCES

You may have noticed that flowers usually don't occur singly, but rather in groups of varying appearance known as **inflorescences**. Here are some of the most common ones:

Male flower

Chestnut branch with inflorescences of male flowers, plus female flowers at the base

Female flower

DIFFERENT TYPES OF INFLORESCENCES

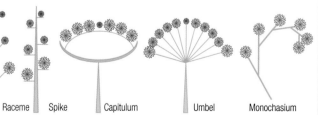

Raceme   Spike   Capitulum   Umbel   Monochasium

Corymb   Panicle   Dichasium

MALES AND FEMALES AMONG THE PLANTS

If you walk through the woods and see **holly** plants loaded with red fruits next to others of the same type that have no fruit, you are standing in front of a fine example of a **dioecious** plant. The specimens that have no fruit are male; they have only male flowers. Only the females, with female flowers, can bear fruit.

FRUITS

When you hear the word *fruit* you probably think of the sweet, fleshy fruits that we commonly eat as snacks and desserts, such as pears, grapes, strawberries, and peaches. But for a botanist, fruits also include nuts, the pods of peas and beans, kernels of corn, peppers, and even the winged *seeds* of the maple and ash trees along town and city walkways. A **fruit** is the **ovary** of the developed flower, inside which the **seeds** are located.

HOW IS A FRUIT FORMED?

When **pollination** occurs, followed by **fertilization**, the **carpels** that make up the ovary of the flower begin a new stage of development: they grow, change shape, harden or turn fleshy, and little by little change into a mature fruit. From the outside toward the inside, almost all fruits consist of an outer layer (the **epicarp**) and an inner one (the **endocarp**), which surrounds the seeds. Many of them also have an intermediate layer known as the **mesocarp**, which usually is fleshy.

The part of a peach that we eat is the mesocarp; the seed is the inner part (the pit). In contrast, the part of a chestnut that we eat is the seed.

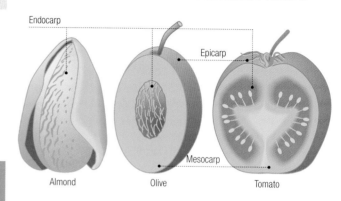

Endocarp

Epicarp

Mesocarp

Almond   Olive   Tomato

Various positions of the ovary in flowers

SEEDLESS FRUITS

It may not seem to make much sense, but in some plants, the mere pollination of the stigma causes the fruit to begin developing without fertilization. Naturally, no seeds are formed that way. The seedless grape and the banana are examples of a phenomenon known by the name of **parthenocarpy**.

THE ROLE OF FRUITS IN NATURE

Fruits can protect the seeds they contain for a certain time, but even more importantly, they are a specialized structured for **dispersing seeds**. **Dry fruits** commonly are adapted to being transported by the wind and by animals (stuck to their fur or feathers) and may be **dehiscent**, if they open to release seeds when they are mature, or **indehiscent**, if they never open and have to decompose to release their seeds. **Fleshy fruits** are eaten by animals, and their seeds travel in their stomachs to be expelled along with the feces in some other place.

A SINGLE FRUIT OR MANY FRUITS TOGETHER?

Among the fruits that you know, there are some that come from a flower with a single carpel. These are known as **simple fruits**, such as the date, the coconut, and the cherry. Others, known as **aggregate fruits**, originated from a flower with several pistils, such as the blackberry and the raspberry. Finally, the **multiple fruits**, such as the pineapple and the fig, come from the set of flowers in an inflorescence.

When birds eat fruits in one place and release the seeds in other places, sometimes very far away, they perform an important role in dispersion.

Cherries are simple fruits.

DRY FRUITS

Dehiscent dry fruits possess special mechanisms for opening and releasing seeds, whether through a ventral suture, through the juncture of the neighboring carpels, through the median nerve of the carpels, and so forth. This type includes the **legumes**, the **siliques**, and the many variants of **capsules**. On the other hand, the **indehiscent dry fruits** lack these mechanisms. If they are simple, they go by the generic name of **nuts**. Others are the **samara**, the **glans**, and the **caryopsis.**

DEHISCENT DRY FRUITS

Bean legume

Poppy capsule

Pimpernel pyxis

Loculicide capsule

INDEHISCENT DRY FRUITS

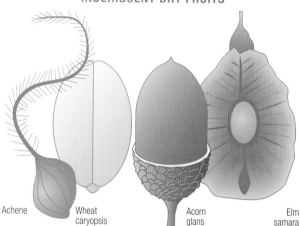

Achene

Wheat caryopsis

Acorn glans

Elm samara

The seeds of apricots, peaches, and other fruits with a pit commonly contain toxic substances, so they should never be eaten.

FALSE BUT EXQUISITE FRUITS

The true fruit of the apple tree is the apple core that surrounds the seeds. The rest of the apple pulp comes from the swelling of the flower's thalamus, which surrounds the ovary until it combines with it. The juicy and bittersweet pulp of the strawberry also is the fleshy **thalamus** on which the many tiny seeds form, and they are the true fruits.

THE PARTS OF A STRAWBERRY

Style

Stamen

Residue of style

Fleshy part

Sepal

Ovary

Sepal

FLESHY FRUITS

The fleshy fruits are indehiscent, and, in general, they have a thick and juicy **mesocarp**. The **epicarp** commonly is thin (the peel), and the **endocarp** may be woody and fairly thick, as with drupes, or else it too is fleshy, as with **berries**. The **watermelon** is a fruit with highly developed placentas that extend from the center to the carpelar wall.

FLESHY FRUITS

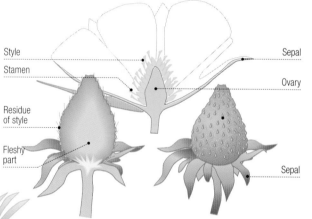

Tomato berry

Pineapple sorosis

Apricot drupe

Fig syconium

AN EXCEPTIONAL FRUIT

You may find it strange that the edible part of one of the most highly esteemed fruits in the whole world, the orange, is not precisely the mesocarp, which, in this case, is the white, spongy layer that you peel away with the thin orange epicarp. The sections are the membranous endocarpal lining of vesicles filled with juice.

SEEDS

You can imagine the seed as a young, undeveloped plant that is living but in a temporary state of rest; provided with a reserve of nutritive substances, and generally surrounded by a protective covering—and sometimes by a fruit—it is ready for dispersion. When the young plant begins to grow, during the first days, it takes nourishment from the reserves that accompany it.

HOW ARE SEEDS FORMED?

Not all plants produce seeds. In plants that are adapted to land environments, at the end of fertilization, the **ovule** and its **embryo** turn into a **seed**; its walls thicken and form the outer covering or **episperm**, and the embryo remains surrounded by a nutritive tissue known as **endosperm**.

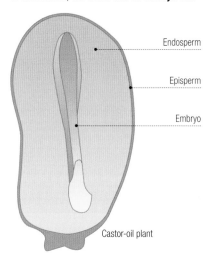

Endosperm

Episperm

Embryo

Castor-oil plant

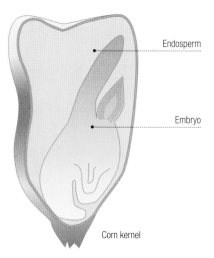

Endosperm

Embryo

Corn kernel

Bean seed (the fleshy cotyledons fulfill the function of the endosperm)

THE ADVANTAGES OF HAVING SEEDS

The seed is like an invention for protecting the young plantlet in its first stages of development. On the one hand, the episperm protects the delicate embryo from many parasites, drying out, excessive heat and cold, mechanical damage, and the chemical action of the digestive juices of animals. On the other hand, when the new plant begins its growth, it takes nourishment from the endosperm until it is capable of living on its own.

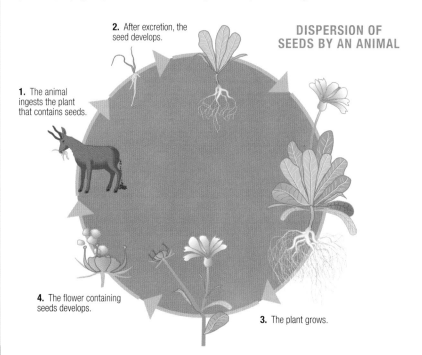

2. After excretion, the seed develops.

1. The animal ingests the plant that contains seeds.

DISPERSION OF SEEDS BY AN ANIMAL

4. The flower containing seeds develops.

3. The plant grows.

Heads or ears of wheat

THE IMPORTANCE OF ENDOSPERM FOR HUMANITY

Endosperm is more than a nutritional reserve for the embryo of many plants; it serves the same function for a large portion of the human population. Human beings cultivate many plants for the food value of the endosperm in their seeds. The grain **wheat** and other cereals, for example, are made up mostly of the endosperm of the seed, which in this case is farinaceous.

SLEEPING FOR SURVIVAL

What would happen if the seeds of a desert plant germinated when they fell? Undoubtedly, the new plants would dry out very quickly. But if these seeds remain in a lethargic state until some day when a rain "awakens" them, the newborn plants will have a much better chance of surviving and growing. The lethargy of the seeds is thus a survival strategy.

Depending on the plant's lifestyle, every species requires different conditions for ending the dormancy of the seeds.

Seed waiting for favorable conditions

After a rainfall, it sprouts, grows . . .

produces seeds . . .

and dies in a matter of weeks

SEEDS THAT "WAKE UP" WITH FIRE

Some seeds germinate only after a fire. This happens because the fire helps to crack the seed open or, in the case of pinecones, because the seed is released after the fire melts a waxy pitch contained in the cone. The advantage of depending on fire for germination is that the flames will destroy all competing vegetation around the seed.

The seeds of the rockrose are very resistant to fire.

PLANTS THAT SOW THEIR SEEDS

Once the flowers of the **peanut** plant have been fertilized, the stalks bend toward the ground and insert the ovaries into the soil. That way the fruits (the peanuts) mature underground.

TRAVELING IN THE SHAPE OF A SEED

Unlike animals, plants can't move on their own, but they can still travel. When they reach maturity, the dehiscent fruits of many plants, such as legumes, burst open on dry, hot days, casting their seeds afar. Winged seeds and seeds that are covered with hairs are carried by the wind. Others travel stuck to the fur or feathers of animals or are carried by water currents. Finally, certain seeds are dispersed by animals that are incapable of digesting them once they have devoured the fruits that contain them.

Example of seeds dispersed by the wind

Seeds from the **indigo lotus** more than a thousand years old have been made to germinate by abrading them before moistening them.

GERMINATION

After a dormancy that lasts for different times depending on the species, the embryo wakes up when the temperature and moisture conditions are right for growth. Then the seed swells, and the embryo begins to grow by using its nutritional reserves and breaks through the episperm. When the reserves are all used up, the young plant already has a root with root hairs so it can absorb nutrients from the soil on it own and the first green leaves with chlorophyll for carrying out photosynthesis.

PHYSICAL CONDITIONS OF THE ENVIRONMENT

Every plant species needs certain materials and specific environmental conditions for growth and reproduction. Whether or not we find a plant in a certain region depends on moisture, light, temperature, the type of soil, and other physical factors, as well as the presence or absence of other competing plants that are better adapted.

LIGHT

Sunlight is the basis of existence for photosynthetic plants, and its distribution is a function of how cloudy the area is and how long the clouds are present, as well as the type of vegetation.

CHECK THE IMPORTANCE OF LIGHT

If you plant one bean seed in each of two pots and place one of them in a dark environment, you will see how it grows long and pale; in short, it becomes weak. The plant devotes all its energy toward a single goal: getting out of the dark area. If it doesn't succeed, it dies. Without light, photosynthesis is impossible.

Three-week old bean plants: on the left, the one kept in darkness is thin and weak; the one on the right has grown in the light.

Less light penetrates into a spruce and birch forest (right) than into a pine forest (left).

PLANTS FOR SHORT AND LONG DAYS

The blooming of some plants depends on the length of the day, that is, the number of hours of light per day, which is known as the **photoperiod**. There are plants, such as the chrysanthemums, that produce no flowers when the photoperiod exceeds a certain number of hours, and, as a result, they commonly bloom only in the fall; these are called **short-day species**. On the other hand, **long-day** species, such as the **gladiolus**, need a minimum number of hours in the light to bloom, so they bloom in the spring and summer. There are also plants whose blooming is not affected by the length of the day.

Ivy blooms only if it is exposed to direct light and does not bloom in the shade.

THE BEHAVIOR OF CERTAIN PLANTS DEPENDING ON THE LENGTH OF THE DAY

Long day plants	Short day plants	Average day plants
Wheat	Rice	Kentucky bluegrass
Barley	Millet	Candytuft
Peas	Hemp	Chickweed
Mustard	Soy	Common groundsel
Spinach	Chrysanthemum	Cucumber
Grapevine	Poinsettia	Tomato
Red clover	White clover	Carrot
Bean	Amaranth	

The poinsettia is a plant that blooms in December, when the days are short; however, you can make it bloom at any time of the year if you artificially control the length of the day.

TEMPERATURE AND MOISTURE

Heat is a form of energy that comes mainly from the sun's rays, and it is expressed in terms of temperature. Within certain limits, a rise in temperature stimulates plant growth; but very high or low temperatures retard growth. Plants are adapted to the areas where they live, especially in relation to the **climate**, which depends mainly on temperature and precipitation.

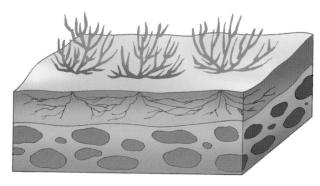

Some desert plants have interconnected roots to take better advantage of the little moisture in the subsoil.

SOIL

The soil is another important factor in the life of land plants, because it contains the water and the minerals that these plants use for nutrients. There are many types of soils, and there also are plants that are better adapted than others to each soil type. A chestnut, for example, cannot develop in a soil where the bedrock is very close to the surface. On the other hand, there are plants that grow in fissures of rocks and lichens that cover their surface like a carpet. Some plants thrive in salty soils, and others like limestone soils, and so forth.

The temperature and humidity inside greenhouses are controlled artificially to imitate the conditions to which the plants being cultivated are adapted. That way it is possible to produce certain vegetables and flowers at any time of the year, even with summer plants.

MANY ROOTS AND FEW LEAVES

You probably have noticed that in the desert the plants seem to be quite far from one another. In fact they are in contact through their extensive underground organs. If they developed their aerial parts very extensively, they would perish through evaporation. This is in total contrast to what happens with plants that live in swampy or aquatic environments, where there is no worry about water and a plant can afford the luxury of having lots of leaves.

Some forest trees in Siberia withstand temperatures of 50°F below (−46°C). Some microscopic algae in the diatom group can withstand temperatures as low as −328°F (−200°C)!

A mature soil consists of three layers that are known as **horizons A, B, and C**; together they constitute the **soil profile**. The upper horizon (A), which is the richest in **humus**, is found on top of a subsoil (B), which contains lots of minerals. Beneath the subsoil is the bedrock, which is in the process of decay (C).

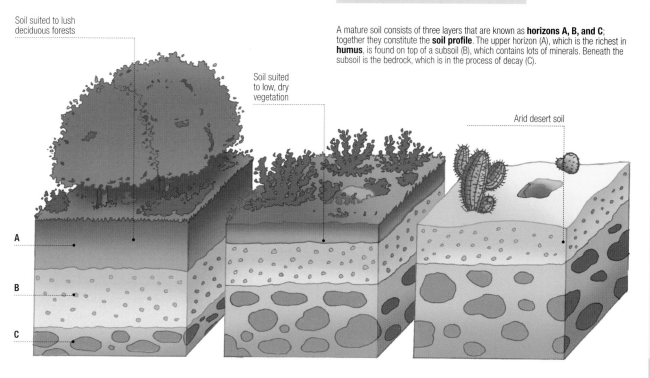

Soil suited to lush deciduous forests

Soil suited to low, dry vegetation

Arid desert soil

A

B

C

PLANT COMMUNITIES AND ECOSYSTEMS

We can study what a plant or an animal, or a group of individuals, is like and what makes it that way; but its life really depends strictly on the physical environment that surrounds it and on the other living beings with which it shares the territory where it lives. **Ecology** involves studying the relationships among living beings and their physical environment, how they influence one another, and how every individual interacts with the others.

WHAT IS A SPECIES?

The different types of plants that are found in a forest, such as red oaks, pines, rhododendron, fern, and others, are obviously the different plant species that make up the forest. But if you look more closely, you will see that there are two classes of pines: some that are taller and have denser crowns (piñon pines) and others that exhibit more extensive branching. That means that there are two species of pine. The individuals of one species reproduce with one another; individuals from different species cannot produce fertile offspring.

VARIETIES AND BREEDS

If you go to buy apples, you have to choose among several varieties. They come in different colors, shapes, and tastes. In addition, not all apples ripen at the same time. These are varieties (or breeds) of a single species: the apple tree.

The plant community is made up of:
The population of red oaks (species) and the individual (red oak tree); the population of piñon pines (species) and the individual (piñon tree); the population of blackhaw viburnums (species) and the individual (blackhaw viburnums plant); the population of English hawthorn (species) and the individual English hawthorn, etc.

The animal community is made up of:
The population of eagles (species) and the individual eagle; the population of wild pigs (species) and the individual wild pig, and so forth.

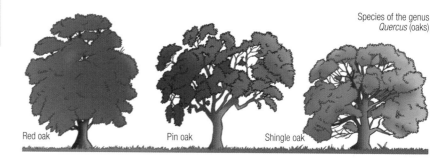

How many species a community has indicates its **biodiversity**.

Species from the *Pinus* (pine) genus

Piñon pine

White pine

Austrian pine

Species of the genus *Quercus* (oaks)

Red oak

Pin oak

Shingle oak

THE PLANT COMMUNITY

In nature, every species is represented by a **population** of individuals that reproduce among themselves. For example, all the piñon pines in a forest make up the population of piñon pines in that forest. All the plant populations of a given place make up the **plant community** that inhabits it. If we also consider the animal populations, we have the community of living beings of that place, that is, the **biotic community**.

Most of the terrain that is not covered by forests today is populated by plant communities that owe their existence to human beings.

Mediterranean coast populated by pines

Introduction

Plant Anatomy

Plant Physiology

Reproduction

Flower, Fruit, and Seed

Ecology and Evolution

Algae

Fungi

Plants

Plants with Flowers and Fruits

Plants and Their Environment

Aquatic Plants

Wild Plants

Domesticated Plants

Gardens

Alphabetical Subject Index

ECOSYSTEMS

Together the biotic community (or **biocenosis**) and the physical environment (soil, climate, etc.) make up the **ecosystem**. All the physical and biological components of an ecosystem are interdependent, and there is a constant exchange of materials and energy among them.

A small pond is one example of an ecosystem. The plant community is integrated by populations of different riparian and aquatic species, plus multicelled and microscopic algae.

The stratification of plant life in a forest reflects the different niches in which every plant form meets its vital needs.

Arboreal stratum

Herbaceous stratum

Bushy stratum

Edaphic stratum

Muscineal stratum

THE ECOLOGICAL PYRAMID

In every ecosystem, the herbivores feed on algae or plants, and the carnivores eat the herbivores or other carnivores. The scavengers take care of carcasses, and the bacteria and decomposing fungi mineralize the organic remains so they can be used by plants. This is the cycle of matter and energy.

Matter is constantly recycled, but energy is lost in each of these steps or links in the chain. If life continues, it's because algae and plants "function" with solar energy; they are the real **producers** in nature. All other organisms are **consumers**.

Sample ecological pyramid of an ecosystem

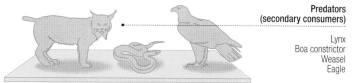

Scavengers

Black buzzard

Predators (secondary consumers)

Lynx
Boa constrictor
Weasel
Eagle

In the long run, there are never two species that occupy the same niche in an ecosystem, since they would be in continual competition. One always succeeds in displacing the other.

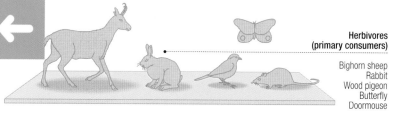

Herbivores (primary consumers)

Bighorn sheep
Rabbit
Wood pigeon
Butterfly
Doormouse

ECOLOGICAL NICHE

The fact that we find different species in the same place is because of the different lifestyle of each one and their different use of the environment with respect to other species in the community. This way of using environmental conditions (light, food, space, etc.) is known as an ecological niche.

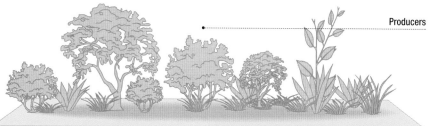

Producers

ENDEMISMS AND FLORAL KINGDOMS

It's possible to observe the plant covering of any given area from two different perspectives. If you try to identify every one of the species present to "take inventory," you will be studying the **flora** of that area. On the other hand, if you are interested in the physiognomy of the plants and the countryside, if they make up a forest, a field, or a thicket, for example, then you will be studying the **vegetation**.

HOW IS AN ENDEMISM FORMED?

A species that is found in just one area when it could be found in others, based on its physiological characteristics, is said to be **endemic** to that terrain, or that it is an **endemism**. An endemism can be produced in two ways: by formation of a new species that remains isolated from its former colleagues, or because a species that was widespread in earlier times is now isolated in a restricted area.

You will find a large number of endemisms on the summits of high mountains and on islands far away from the continents.

Climatic changes   Thousands of years later

In progressive endemism, a group of individuals of one species remains isolated and differentiates progressively from others of its type because it lives under different conditions.

Immigration of herbivore consumers   Extinction

In conservative endemism, the old members of the population die off, and the group that has remained isolated from the causes of the extinction survive.

THE MAIN CAUSES FOR THE EXTINCTION OF SPECIES

Natural Causes	Major climatic changes
	Diseases and epidemics
	Unequal ability to compete
	Loss of reproductive capacity
Causes of Human Origin	Extensive agriculture
	Deforestation and plowing
	Excessive grazing
	Industrialization and urbanization
	Large public works
	Mining
	Forest fires
	Genetic contamination
	Uncontrolled harvesting
	Lack of pollinators because of insecticide abuse

CAUSES OF ENDEMISM THROUGH ISOLATION

- **Mountains** are ecological islands because they are separated from the valleys that surround them by different climatic conditions.

- **Deserts** also are islands because of the hostility of their climate.

- **Special soils**, such as a soil that's high in gypsum, are very selective, isolated areas for plants.

- **Islands** surrounded by the sea are the areas where endemism appears with the greatest potential for land plants.

GEOGRAPHICAL AND ECOLOGICAL BARRIERS

The main cause of the formation of endemisms is the isolation of the population through barriers that inhibit expansion. These barriers may be geographical but also ecological, in other words, because of major differences in significant ecological or seasonal factors. This last case occurs when there is no coincidence in the pollination season.

Incahuasi is a peculiar island in Bolivia; it is surrounded by a salt lake, and many species of cactus have become indigenous to it.

At the present time, there are more than 20,000 plant species in the world that are in **danger of extinction**.

THE FLORAL KINGDOMS OF THE PLANET

Just as the type of vegetation that you find in any given place is mostly because of the present climate and soil conditions in the area, the **flora** (the species) is the result of past happenings on our plant that produced endimisms on a continental scale. For example, the tropical rain forests of South America have the same type of vegetation as the ones in Africa and Asia, but the species are not the same.

There are six major **floral kingdoms** that are distinguished in the world; each has different characteristics relative to the flora.

☐ Holarctic kingdom	☐ Neotropical kingdom	☐ Australian kingdom
☐ Paleotropical kingdom	☐ Cape kingdom	☐ Antarctic kingdom

THE HISTORY OF THE CONTINENTS

230 million years ago

200 million years ago

135 million years ago

60 million years ago

Rockrose, a flammable shrub of the Mediterranean maquis.

THE HISTORY OF THE CONTINENTS

The surface of our planet is made up of a series of **lithospheric plates** that have not always been distributed the way they currently are. Some continents individualized in very remote times, isolating many species that later evolved separately, but others remained connected until not long ago, in terms of millions of years. In addition, the climate has undergone many changes throughout the history of the earth, in such a way that areas that today are covered with ice once were covered by forests and vice versa.

VICARIANT SPECIES

When botanists travel on other continents, they are most interested in finding species that carry out the same function in the plant community as the ones they know in different but similar environments. These are called **vicariant species**. For example, in the laurel underbrush of Chile and the **chaparral** of California, the **rockrose** of the **Mediterranean maquis** has two vicariant species: the **chamise** and the **red shanks**, respectively. The three species occupy the same type of **ecological niche**: they are "flammable" bushes that encourage fire to displace other species. Their seeds are resistant to fire.

VEGETATION AND LANDSCAPE

When you look at a **landscape**, one of its main components is the vegetation. Humans have eliminated many forests and other **plant formations**; but, fortunately, the plants frequently return to colonize the territory. In other places, the damage may, unfortunately, be irreparable.

TYPES OF PLANTS

Two or more plant species are of the same type if they fulfill the same function in the community. This function is closely linked to the degree of protection that the plant provides for its organs of regeneration (seeds, buds, etc.) during the hostile season (cold or dryness). And that is directly related to the height at which those organs are produced.

TYPES OF PLANTS ACCORDING TO THEIR ORGANS OF REGENERATION

    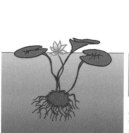

In **trees** and **shrubs** the buds for renewal are more than one foot (30 cm) above the ground.

Dwarf shrubs are plants with a woody and resistant lower part, with buds less than one foot (30 cm) above the ground.

Rosette plants have their regenerative organs at ground level.

Plants with tubers, rhizomes, or **bulbs** have their organs underground.

Aquatic plants have regenerative buds under water.

Annual grasses have only one organ of regeneration, the seed.

TYPES OF VEGETATION

A given type of vegetation, or **plant formation**, is not characterized by the species that make it up, but instead by the type of species and the proportion in which they are combined—in other words, by the physiognomy of the whole, which is one of the main factors that determine the **landscape.**

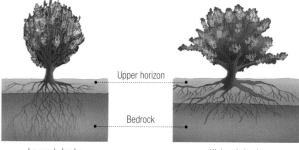

Upper horizon

Bedrock

Low red shanks
(*Adenostoma sparsifolium*)

High red shanks
(*Adenostoma sparsifolium*)

PRINCIPAL PLANT FORMATIONS

Forests and woods	Trees predominate.
Savannahs and pasturelands	Trees form an open landscape and grasses thrive among them.
Thickets	Shrubs predominate.
Steppes and prairies	Grasses predominate.

An extensive forest that is wild instead of cultivated and contains abundant vegetation is known as a jungle.

The water lily is an aquatic plant whose leaves float on the water.

PLANTS LIVING TOGETHER

Sometimes you may wonder how two very similar plants can live together without competing with one another. For example, in the chaparral of California, there are two species of **red shanks** (*Adenostoma sparsifolium*) that live next to one another, but they use different resources. The lower one is rooted in the bedrock of the soil; the other roots in the upper horizon. The first flowers and produces seeds at the end of the spring; the second, at the end of the summer. As you can see, competition is minimal.

SUCCESSION AND EQUILIBRIUM

Even though the climatic conditions and type of soil determine the vegetation of any location, the vegetation, in its turn, has an influence on the soil and, on a small scale, on the climate. These interactions lead to a **state of equilibrium** or **climax** between the vegetation, the climate, and the soil of the area. Any change, whether natural or artificial, in one of them catalyzes an **ecological succession** that tends to reestablish the balance, which is embodied in the plants referred to as the **climax vegetation.**

Spruce forests

The spruces displace the pines, which cannot thrive in the shade; 500 years after the fire, the original spruce forest, the climax community, has reestablished itself.

In 150 years, white pines displace the birches; spruces can thrive in the shade of the pines.

Example of ecological succession after a fire in a spruce forest in Northern Europe.

Fire

As early as the first year, a pasture of light-loving grasses sprouts.

Soon birch and poplar shoots appear; they need light and grow quickly.

At the end of 60 years, a birch forest has formed, in the shelter of which the white pine grows.

THE EARTH'S CLIMATES AND VEGETATION

Throughout the world, there are regions that are characterized by their climate, based on the annual precipitation, the average annual temperature, and the relationship between rainfall and evaporation. With this in mind, it's possible to predict what kind of landscape to expect in each of these regions. In gross terms, the climate gets colder the farther we go from the equator; but, on high mountains, the temperature also falls as we climb higher.

VEGETATION ZONES

If you climb a mountain, you notice that the higher you go, the colder it gets. Plants also notice this; that's why the vegetation changes with altitude. In the Pyrenees Mountains between France and Spain, for example, there are four identifiable **vegetation zones**:

• **the basal zone**: up to 2,700 feet (900 m) in altitude. The characteristic landscape is the holm oak forest;

• **the montane zone**: from 2,700 to 5,300 feet (900 to 1,700 m). Forests of beech, oak, pine, and birch predominate;

• **the subalpine zone**: between 5,300 and 6,800 feet (1,700 and 2,200 m). The black pine dominates;

• **the alpine zone**: from 6,800 feet (2,200 m) to the clouds. There are no more trees, and areas of perennial grasses predominate.

THE DIFFERENT TYPES OF VEGETATION OF THE VARIOUS CLIMATIC REGIONS ON EARTH

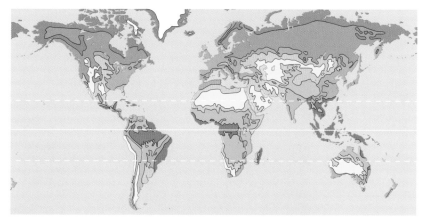

You can follow the steps of a **succession** by observing year after year how an abandoned farm field evolves. Eventually, it will return to the forest that it once was.

In a **succession**, all the communities except for the climax vegetation **die of success**, because it is their success that creates the favorable conditions for establishing the subsequent community.

Tropical rain forest	Temperate deciduous forest because of cold	Steppes and grasslands	Tundra
Temperate moist forests	Mediterranean forests and scrub	Deserts and semideserts	Ice
Tropical and subtropical deciduous forests because of drought	Conifer forests	High-mountain vegetation	

Introduction

Plant Anatomy

Plant Physiology

Reproduction

Flower, Fruit, and Seed

Ecology and Evolution

Algae

Fungi

Plants

Plants with Flowers and Fruits

Plants and Their Environment

Aquatic Plants

Wild Plants

Domesticated Plants

Gardens

Alphabetical Subject Index

SPECIAL FORMS OF PLANT LIFE

The form of plant life that everyone knows is a plant with a stem, leaves, and roots, which takes nourishment from the minerals in the soil by carrying out photosynthesis. But there are many plants that live differently, beginning with the fungi, which have no chlorophyll, and ending with plants that have chlorophyll that have developed a strategy for "saving work" at the expense of others.

DECOMPOSING DEAD BODIES TO EAT

Saprophytes are organisms that are incapable of either producing their own foods or ingesting solid foods. They use the **enzymes** they produce to dissolve the organic substances of plant and animal **cadavers**, plus their offal, and then they absorb them directly through their cell membranes.

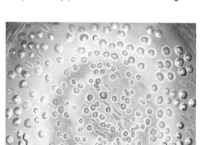

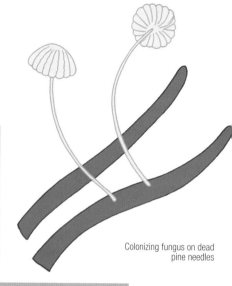

Dead organic material is decomposed and putrefied by the action of bacteria, yeasts (in the photo), and fungi that live as saprophytes.

In a field of lupines the size of a football field, these plants can use more than 450 pounds (200 kg) of nitrogen fixed by the bacteria in their roots.

Colonizing fungus on dead pine needles

MUTUAL AID

Symbiosis is an association between two living beings of different species, known as **symbionts**, in which both organisms obtain some benefit, frequently of a nutritional kind. Sometimes the association is very close and permanent, as with **lichens** (the association of an alga and a fungus) and **nitrogen fixation** (an association between the roots of a leguminous plant and certain bacteria that fix the free nitrogen from the air).

Moss, which often covers part of the bark of trees in moist areas, uses the trees as a substrate.

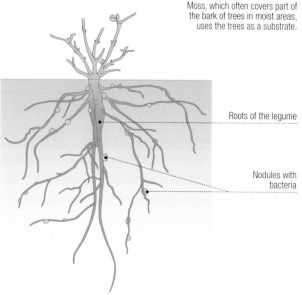

Roots of the legume

Nodules with bacteria

Nitrogen fixation. The bacteria penetrate into the root cells and develop and multiply by using the plant substances. Before the "infection," the cells of the root multiply, increase in size, produce nodules, and take advantage of the nitrogen fixed by the bacteria.

ONE MEMBER BENEFITS AND NO ONE IS HARMED

This is the case of the epiphytic plants that live on other plants without harming them. The fern known as the elkhorn fern, for example, forms a medley of roots in a crack of a tree's bark, where ants make a nest and accumulate humus rich in nutrients for the fern. This type of relationship is known as **commensalism**.

LIVING AT THE EXPENSE OF OTHERS

A parasite is a living being that gets its nourishment at the expense of another living creature, which is known as the **host**, by living on it (an **ectoparasite**) or inside its body (an **endoparasite**). The only difference between a predator and a parasite is that the latter doesn't kill its victim to devour it, but instead uses it while it's alive. Parasites cause diseases, because they destroy the host's cells or produce toxic substances.

For the Gallic druids, if mistletoe grew on an oak, it was considered to be sent from the heavens, and the tree was believed to be chosen by God.

Complete parasites have no chlorophyll, and they, therefore, have to get all the necessary nutrients from the host. As a result, they hook in directly to the host's phloem vessels, which conduct elaborated sap. The drawing shows a willow branch parasitized by common dodder.

Clover roots parasitized by sunflower broomrape (*Orobanche cumana*).

Branch of an apple tree parasitized by mistletoe. Mistletoe is an example of a semiparasite. It has chlorophyll, even in the winter after the apple tree has lost its leaves, that way it can carry out photosynthesis, and it extracts raw sap from the host's xylem only through its haustoria.

HAUSTORIA

Many parasitic fungi, such as the one that causes potato rot, produce an enzyme at the end of their hyphae that helps them penetrate the plant's tissues. Once they are inside, the hyphae extend among the cells and penetrate them by means of haustoria which facilitate the absorption of the substances contained in the cytoplasm.

Fungus

Host cell

Perforation

Ramified haustorium

Above, a fungus penetrating the epidermis of a leaf and forming a haustorium (shown in cross section at right).

THESE ARE NOT FRUITS, BUT GALLS

On the tender leaves and branches of many plants, especially oaks, we often see small deformities or swellings that at first glance may look like fruits. But they are galls. If you open them, you will see that they contain the larvae of a parasitic insect. There are many types of galls. Pine galls are produced by bacteria, as are the warts that appear on olive trees. Other galls are produced by fungi.

Galls on a beech leaf

EVOLUTION IN THE PLANT WORLD

You probably realize that the reptiles that are alive today are not like the dinosaurs that lived millions of years ago, and the plants of our landscape are not like the ones that fed the gigantic diplodocus. Ever since life first appeared on our planet, all living beings have been in a continual process of change.

Variability among descendents is tested every time there are changes in the environment, and the individuals that are most prepared for life under the new conditions are the ones that survive. This process of **natural selection** has been the motive force for the **evolution** of life.

THE OLDEST ORGANISMS

Among the living organisms that inhabited our planet were **cyanobacteria** similar to the present-day blue-green algae. These organisms contributed to the transformation from the primitive atmosphere of the earth, which had no oxygen, to an atmosphere like the current one, with oxygen and a protective ozone layer to cut down on the sun's ultraviolet radiation.

THE WATERS FILL WITH LIFE

After the first bacterial cells, many other groups of algae evolved, first one-celled and later on multicelled ones, with their **genetic material (DNA)** organized into **chromosomes** stuffed inside a nucleus. These algae could come closer to the surface of the water and invade the damp coasts.

THE GREAT INVENTION OF SEXUAL REPRODUCTION

The first algae reproduced by dividing to create two cells identical to the mother cell. The algae with a nucleus began a new system: two cells got together, exchanged a part of their DNA, and divided. The descendents contained a mixture of their parents' DNA and were no longer identical. This **variability** created more possibilities for adaptation and sped up **evolution**, producing an explosion of life forms.

THE EVOLUTIONARY TREE OF LIFE

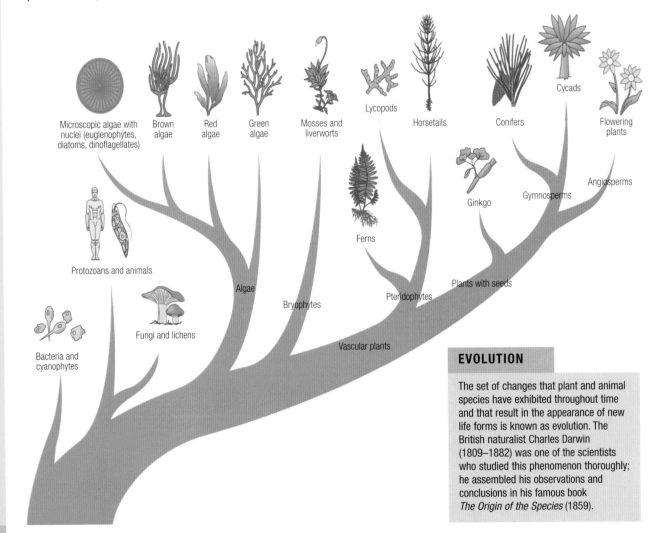

Microscopic algae with nuclei (euglenophytes, diatoms, dinoflagellates)
Brown algae
Red algae
Green algae
Mosses and liverworts
Lycopods
Horsetails
Conifers
Cycads
Flowering plants
Angiosperms
Ginkgo
Gymnosperms
Ferns
Protozoans and animals
Algae
Bryophytes
Plants with seeds
Pteridophytes
Fungi and lichens
Vascular plants
Bacteria and cyanophytes

EVOLUTION

The set of changes that plant and animal species have exhibited throughout time and that result in the appearance of new life forms is known as evolution. The British naturalist Charles Darwin (1809–1882) was one of the scientists who studied this phenomenon thoroughly; he assembled his observations and conclusions in his famous book *The Origin of the Species* (1859).

THE FIRST LANDING

For millions of years there was no life on the exposed land—not until some **green algae** (**chlorophytes**) on the shores of lakes and swamps developed a waxy surface, the **cuticle**, which prevented desiccation when the water level went down. Some tiny openings, the **stomata**, allowed the entry of the carbon dioxide necessary for photosynthesis and the exit of oxygen. They were like the present-day **mosses** and **liverworts**, which are terrestrial but have to remain in moist, dark environments because they disseminate gametes that have to travel to meet one another.

GAINING HEIGHT

After the success of the vascular plants, vegetation became denser. In order to get light, plants had to grow taller than their neighbors and that involved additional support. That gave rise to the **woody tissue** that made it possible for the **first trees** to appear.

NEW FOODS, NEW APPEARANCES

With the development of terrestrial plant life, plant remains began accumulating for the first time; that stimulated the development of saprophytic **fungi** from algae that lost their chlorophyll. These fungi fed on the dead matter, and, as they decomposed it, they began forming the **first fertile soil** in which the first vascular plants sank their roots.

Thanks to their great adaptability, mosses are found in all land environments. They are capable of retaining lots of moisture and resisting drought for a long time.

Fifty-five million years ago the earth's climate was so warm that tropical vegetation reached the polar circles.

THE SECOND LANDING

A second group of algae followed a different evolutionary path from that of the primitive mosses: it found environments for joining gametes without dispersing them in water and developed roots and efficient systems for circulating water. These are the **vascular plants** that now dominate land environments, although they don't all have the same degree of efficiency achieved by the **plants with seeds**, the most highly evolved of all terrestrial plants.

Flowers appeared later. They were a strategy for attracting pollinating insects and birds to help spread their species.

One of the first trees that cast a shadow on the earth's surface was the *Glossopteris*, which existed in great numbers 300 million years ago.

The leaves of plants from places with climatic seasons are more edible than the ones from plants that grow where there are no seasons, since it doesn't make much sense for the former to emphasize precautions against the voracity of animals.

MICROSCOPIC ALGAE

Even though you can't see the **one-celled algae** with the naked eye, you can see the color that they give to the water of a pond or an abandoned pool, or the various color spots that appear on exposed rocks and damp ground, made up of millions of these algae. Like all algae, these are photosynthetic plants that have other pigments besides chlorophyll, which are responsible for the colors they produce.

THE TINIEST AND THE MOST RESISTANT

These are the **cyanobacteria** (or **cyanophytes**), known as the **blue-greenish algae** because they commonly have this color, even though sometimes they are reddish, brown, or almost black. They multiply only asexually, but no other alga or land plant surpasses their ability to occupy environments with extreme conditions of light, cold, heat, or drought. Thanks to the gelatinous cover that they secrete, they can even resist the sun's ultraviolet radiation, which would fry any other living creature.

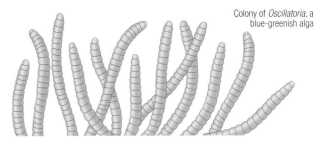

Colony of *Oscillatoria*, a blue-greenish alga

CELLULAR ORGANIZATION OF A NONBACTERIAL MICROSCOPIC ALGA

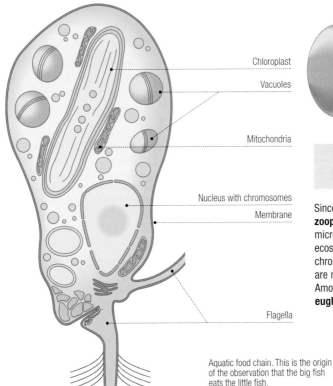

- Chloroplast
- Vacuoles
- Mitochondria
- Nucleus with chromosomes
- Membrane
- Flagella

Aquatic food chain. This is the origin of the observation that the big fish eats the little fish.

THE STRUCTURE OF A BLUE-GREENISH ALGA

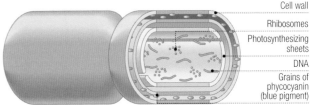

- Cell wall
- Rhibosomes
- Photosynthesizing sheets
- DNA
- Grains of phycocyanin (blue pigment)

THE BASIC FOOD IN SEAS, RIVERS, AND LAKES

Since the protozoa and small aquatic animals that make up the **zooplankton** aren't very fond of the blue-greenish algae, the other microscopic algae are the base of the food chain of all the aquatic ecosystems in the world. These algae have their DNA organized in chromosomes, and, in general, they reproduce sexually. Many of them are mobile, so they used to be referred to as **photosynthetic animals**. Among the best known are the **diatoms**, the **dinoflagellates**, and the **euglenophytes**.

 The coastal waters of certain places can harbor a heightened **phytoplankton** density—up to 375 million individuals per cubic yard (meter) of water!

LIVING AFLOAT

If you study a drop of water from the surface of the sea or from a lake with a microscope, you will see a great number of creatures. These organisms, which live by floating at the mercy of the waves and currents, are **plankton**, made up of algae (**phytoplankton**) and tiny animals (**zooplankton**). Not only zooplankton, but also many animals that feed by filtering water, eat phytoplankton.

BOX-SHAPED CREATURES

The algae known as **diatoms** are found floating freely in water and on moist surfaces. There are even species that live inside the liver and kidneys of humans. Their structure consists of two superimposed halves that fit together like two parts of a box, and their **siliceous walls** have borders, lines, and very fine pores. The remains of their cell walls have built up on the bottom of the oceans over millions of years, and, in some places, they stick out of the surface because of geological lifting. This is what is known as **diatomaceous earth**.

Cyclotella meneghiniana

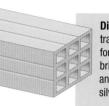

Diatomaceous earth traditionally has been used for making insulating bricks, filters, toothpastes, and powders for polishing silver objects. ⬅

Achnanthes minutissima

⬇ Three quarters of all organic material synthesized in the world and a large part of the atmospheric oxygen are products of the activity of **diatoms** and **dinoflagellates**.

WITH ARMOR AND WHIP

Certain one-celled algae are surrounded by a shell of thick, interlocked **cellulose plates**. They are known as **dinoflagellates** because they have two **flagella** or whips. They have a yellowish, reddish, or brown color because of their large quantity of pigments, and they transmit the color to the water when they form very dense populations. Some species give off a light that is visible on dark nights.

Achnanthes lanceolata

Plant Anatomy

Plant Physiology

Reproduction

Flower, Fruit, and Seed

Ecology and Evolution

Algae

Fungi

Plants

Plants with Flowers and Fruits

Plants and Their Environment

Aquatic Plants

Wild Plants

Domesticated Plants

Gardens

Alphabetical Subject Index

VARIOUS SPECIES OF DINOFLAGELLATES

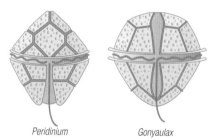

Peridinium   *Gonyaulax*

Dinophysis acuta

Prorocentrum

RED TIDE!

Sometimes right from the beach you can see a large reddish blotch on the surface of the water; this is known as the red tide. It is caused by a great and sudden proliferation of **dinoflagellates**. Red tides are not always toxic, but they may be, and sometimes they even pollute the atmosphere, causing respiratory problems in humans. This is because of the strange way in which algae compete for oxygen: they produce toxic substances.

FOOD POISONING FROM SHELLFISH

You probably have heard of people who have nearly died from eating toxic shellfish. This happens because certain marine animals, such as **mussels**, which feed by filtering water, have ingested toxic substances produced by dinoflagellates. These substances interfere with the diaphragm and cause respiratory failure.

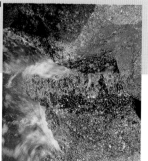

PLANT-ANIMALS

The euglenophyte group contains photosynthetic species, but most of their members lack chlorophyll and are colorless; they live on dead matter or ingest organic particles. Some, like the *Euglena*, which has animal and plant characteristics, have sometimes been classified as a plant (alga) and, at other times, as an animal (protozoan).

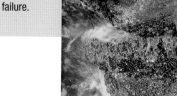

Flagellum
Cytostoma
Stigma
Contractile vacuole

Chloroplasts

Nucleus

Pyrenoid

Membrane

Components of a *Euglena gracilis*.

THE HIGHER ALGAE

When you look at the ocean floor for the first time with underwater goggles, you discover a fascinating world that's very different from the one that we are used to seeing outside the water. The color of this aquatic world is especially caused by the higher algae. Some of them even appear to be plants, but they are not, because, even though they are multicelled, they lack tissues and organs. By living in the water, they need no roots, nor conducting vessels, nor tissues to protect them against desiccation.

A SIMPLE BODY

The higher algae are one-celled, and they generally form colonies of many individuals that live together. The body of the multicelled ones, known as the **thallus**, lacks roots and if it attaches to the substrate, it does so by using **rhizoids**.

Many **additives** used in the food industry come from algae, in particular **stabilizers**, **emulsifiers**, and **thickeners**.

Many algae are cultivated in ponds in the open air to produce vitamins, proteins, and provitamins.

DIFFERENT TYPES OF ORGANIZATION OF HIGHER ALGAE

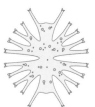

Colony of many independent cells joined by a layer of mucilage

Colony of flagellated individuals that separate only to reproduce

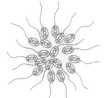

Colony with a fixed number of cells that doesn't vary throughout its life

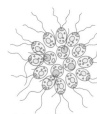

Colony of cells that communicate with one another by bridges, as with *Volvox*

Ramified filament

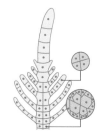

Parenchymatous type

Branched siphonal type

Siphonal type

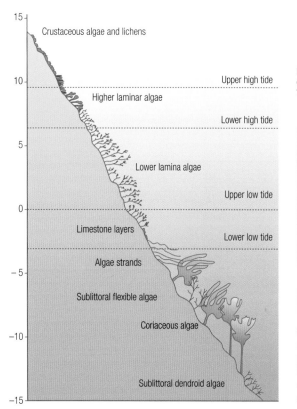

15 —
Crustaceous algae and lichens

10 — Upper high tide
Higher laminar algae
Lower high tide

5 —
Lower lamina algae

Upper low tide

0 —
Limestone layers
Lower low tide

−5 —
Algae strands
Sublittoral flexible algae

Coriaceous algae
−10 —

Sublittoral dendroid algae
−15 —

In an aquarium, eggs and larvae take refuge on algae.

FOOD AND SHELTER FOR AQUATIC ANIMALS

If you have an aquarium at home, you can verify the importance that algae have for fish. In fact, they are important for countless aquatic animals, not just as food and a source of oxygen, but also as a refuge for adults, larvae, and eggs of all kinds. Many fish and crustaceans that are fished where there are no higher algae go to lay their eggs in underwater "forests."

AS FAR AS THE LIGHT PENETRATES

Some algae live outside the water, but they need an aquatic medium for reproduction. However, most of them live in the water wherever the light penetrates. The depth that each alga can reach is related to the minimum **light intensity** that it needs for carrying out **photosynthesis**. Very few algae can survive in the **littoral area** that remains exposed during low tide, because they are exposed to desiccation and extreme temperatures.

GREEN ALGAE

Some green algae are one-celled, but others have filamentous or laminar, green thalluses that look like leaves. In addition to multiplying by fragmentation and spores, they practice several types of sexual reproduction with **alternation of generations**.

Sea lettuce is a green alga that in many places is eaten raw in salads

Brown alga from the genus *Fucus*

BROWN ALGAE

These tend to be brown in color because of a pigment, **fucoxanthin**, that masks the chlorophyll. These are the largest and toughest algae in existence. They may be filamentous, long, thick, and slimy; and they may have thick, ramified sheets, like the *Fucus*.

The worldwide harvest of brown algae is nearly three million tons per year; mostly from plantations in China and Japan.

FLOATING GIANTS

Kelp is floating brown algae of the genus *Macrocystis*, which may reach a length of 70 yards (meters)—longer than many fishing boats. Their mobile reproductive cells contribute greatly to the constant regeneration of the sustaining phytoplankton in every aquatic food chain.

There are two algae that serve as indicators of the quality of seawater: the sea lettuce and *Cystoseira*. The former abounds in polluted areas; the latter can live only in clean waters.

Nori is a red alga from the genus *Porphyra* that the Japanese cultivate for food in the sea.

RED ALGAE

In addition to chlorophyll and phycocyanin, the red algae contain a red pigment known as **phycoerythrin**. These are delicate algae that can't withstand the conditions in places that are subject to tides; that's why they are found in deeper, quiet waters.

Algae are not limited to complementing certain dishes; many algae are used in the pharmaceutical, textile, and even the energy industries (for producing methane).

THE LOWER FUNGI

When we hear about fungi, we imagine an edible mushroom like the ones sold in food stores. But in reality, the fungi are as broad and diverse a kingdom as that of the plants or animals. They include other lower forms that aren't as visible but are equally important from the ecological viewpoint. All fungi are characterized by the fact that they are not photosynthetic. During a long time they have been considered to be plants simply because they **live attached to the soil** or substrate, and, with the exception of the slime molds, they have **rigid cell walls**.

THE SLIME MOLDS

These are the molds that cover the moist surfaces of dead wood, straw, leaf debris, manure, and so forth. They feed by ingesting solid organic particles, spores, bacteria, and other fungi. They produce very resistant **spores**, and, when they germinate, they give rise to **flagellated cells** (myxamebas and swimming cells) that can lead an independent life and that end up functioning like **gametes**. The zygote turns into a multinucleate, mobile **plasmodium** because it has no rigid cell wall; it moves around like a type of slime and feeds by **phagocytosis**.

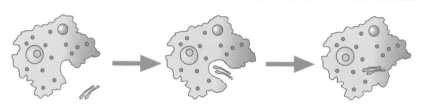

Myxameba phagocytizing bacteria

The most resistant phase in the life cycle of the slime molds is the **spores**. Some can survive more than 75 years!

A WAY TO SURVIVE

Under favorable conditions, the **plasmodia** eat and grow; but if they lack moisture or if the temperature becomes unfavorable, they have their own strategy to survive: the plasmodium changes into a hard, irregular lump known as a **sclerotium**. This allows it to survive up to three years. When favorable conditions return, the sclerotium turns into a plasmodium.

THE LIFE CYCLE OF SLIME MOLDS

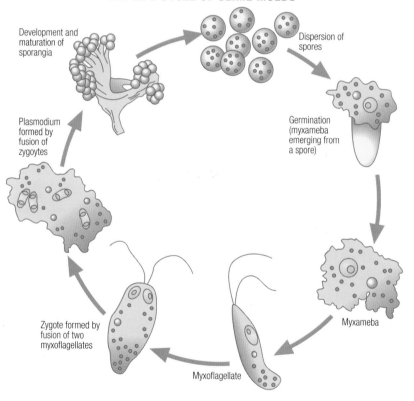

Development and maturation of sporangia

Dispersion of spores

Plasmodium formed by fusion of zygoytes

Germination (myxameba emerging from a spore)

Myxameba

Myxoflagellate

Zygote formed by fusion of two myxoflagellates

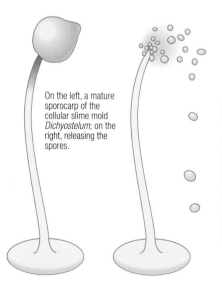

On the left, a mature sporocarp of the cellular slime mold *Dichyostelum*; on the right, releasing the spores.

CELLULAR SLIME MOLDS

In contrast to the true slime molds, these molds don't form a plasmodium, but instead, a type of multicellular body that behaves like a plasmodium (a **pseudoplasmodium**). Some of them resemble a tiny slug that moves at a rate of less than eighty hundredths of an inch (2 mm) per hour. Others are endoparasites of fungi and plants.

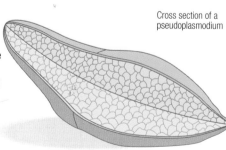

Cross section of a pseudoplasmodium

Introduction

Plant Anatomy

Plant Physiology

Reproduction

Flower, Fruit, and Seed

Ecology and Evolution

Algae

Fungi

Plants

Plants with Flowers and Fruits

Plants and Their Environment

Aquatic Plants

Wild Plants

Domesticated Plants

Gardens

Alphabetical Subject Index

DIGESTING FOOD OUTSIDE THE BODY

All fungi that are not slimy have a strange way of eating. First they break down the food outside their body by using the appropriate chemical substances (**enzymes**), which they produce themselves. They reduce the food to small molecules that can be absorbed through their membrane (**diffusion**), along with any soluble molecules that may be present. This system is known as **lysotrophy**. In order to have lots of surface area in contact with the **substrate** on which they live, most fungi have a body made up of a network of filaments known as **hyphae**, which together make up the **mycelium** of the fungus.

ALGAE-TYPE FUNGI

The simplest lysotrophic fungi resemble algae more than the rest of the fungi, partly because they have a **cell wall of cellulose** instead of chitin. Some are saprophytes, and others are parasites of algae, fungi, aquatic animals, and plants. Many of these fungi have developed a special system for transmitting male nuclei to the females through **copulatory tubes**.

NUTRITION THROUGH EXTERNAL DIGESTION OR LYSOTROPHY

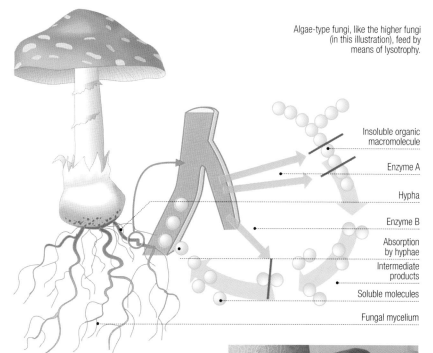

Algae-type fungi, like the higher fungi (in this illustration), feed by means of lysotrophy.

Insoluble organic macromolecule

Enzyme A

Hypha

Enzyme B

Absorption by hyphae

Intermediate products

Soluble molecules

Fungal mycelium

Molds tolerate much higher concentrations of salt and sugar than bacteria do. That's why you may have seen mold in a jar of jelly that has been open for a while.

There are fungi that can withstand very low temperatures, so not even refrigerated foods are safe from them.

DEPENDING ON WATER FOR EATING

You will never see a fungus in a dry place, because, in order to feed, fungi need the presence of water between their hyphae and the substrate. That's the only way the enzymes and soluble products that result from "external digestion" of the substrate can spread out in one direction or another.

It's easy to see fungi in many places in damp woodlots.

THE HIGHER FUNGI

The fungi that are best adapted to life on land, which are known as the higher fungi, are characterized by the presence of **chitin** in their wall, much like the exoskeleton of insects. In addition, these fungi are the only living beings that have a **dicaryotic phase** in their life cycle, in which the cells possess two haploid nuclei (with a single set of chromosomes). They feed by means of **external digestion** and for the most part transform part of their body into a reproductive organ, a **carpophore**, commonly called a **mushroom**.

COLONIZERS OF EXCREMENT

Many fungi that use their hyphae to invade the excrement of herbivorous animals, such as horses and cows, grow very quickly. They carry out sexual reproduction by coming into contact with two **compatible hyphae (zygophores)**, whose extremities swell up and produce **progametangia**. A thin wall delimits a **gametangium** in each one of them. And both gametangia fuse to form a **zygospore** from which a **germinal sporangium** is released.

Given the extremely light weight of spores, it has been calculated that each spore can travel around the world several times transported by the wind before it's deposited in the soil.

The easiest fungi to grow are the saprophytes such as the edible mushrooms, since they grow spontaneously on horse manure.

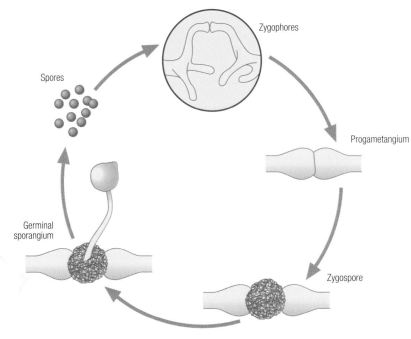

THE LIFE CYCLE OF A FUNGUS OF THE GENUS *MUCOR*, WHICH GROWS ON ANIMAL DUNG

Zygophores

Spores

Progametangium

Germinal sporangium

Zygospore

SAC-TYPE FUNGI

These are the fungi that form their **spores** inside structures shaped like little sacs and known as **asci**; that's why these are known as **ascomycetes**. Their hyphae commonly have **perforated partitions** that permit communication between compartments. In the majority of these fungi, sexual reproduction implies the formation of a type of spores known as **conidia**, which are released from the ends of special hyphae known as **conidiophores**. The conidia, which sometimes are called "summer spores," are a means of quick propagation.

Cross section of an ascocarp (left) and a detail of conidiophore (right).

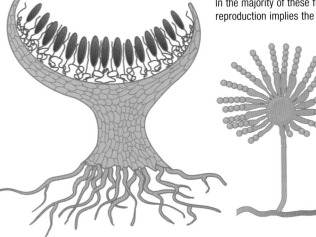

A dog trained for the task, and even a pig, is a good locater of the esteemed truffle, an ascomycetous fungus.

CLUB-TYPE FUNGI

These are the best-known fungi, which form their spores in enlarged hyphal cells in the form of clubs known as **basidia**. The spores (**basidiospores**) appear in fours on the point of the basidium and develop outside instead of inside an ascus. **Mushrooms** are **carpophores** or the fruit-bearing body of the fungus. When they reach maturity, the **cap** opens, and the basidia are found in the **gills** beneath it.

The fungus *Clavaria aurea* or rat's foot

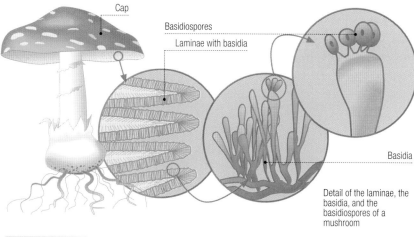

Cap

Basidiospores

Laminae with basidia

Basidia

Detail of the laminae, the basidia, and the basidiospores of a mushroom

Slugs easily eat death cap mushrooms (*Amanita phalloides*). Their ability to withstand the poison of this mushroom is about a thousand times greater than that of humans!

DANGER!

You surely have heard of poisoning from eating toxic mushrooms. People die every year by confusing certain poisonous species with others that are edible. The most toxic of all mushrooms is the death cap (*Amanita phalloides*), also known as the "angel of death." The tasty edible saffron milk cap (*Lactarius deliciosus*) has a "double" (*Lactarius chrysorrheus*) that you can't tell by sight from the true saffron milk cap. This is a mushroom that produces serious digestive troubles. In order to be sure, expert mushroom gatherers scratch the gills (but not with the finger) beneath the cap: a carrot-colored sap runs from the edible mushroom, and a white latex from the poisonous one.

IMPERFECT BUT VERY USEFUL FUNGI

Many fungi, both ascomycetes and basidiomycetes, lack sexual reproduction, so they are known as **imperfect fungi** (or **deuteromycetes**). There are among them several saprophytic species that produce substances that are toxic for certain bacteria and other microbes that compete for the same food. These fungi are grown in laboratories to obtain these substances, which are **antibiotics**, and they are used in combating diseases produced by bacteria. Other species are used in **ripening cheeses** such as Roquefort and Camembert.

In 1928 the British doctor Alexander Fleming observed that a mold had formed on one of his bacteria cultures; it was the deuteromycetes *Penicillium notatum*, which kept the bacteria from growing. Fleming believed that the fungus produced a substance that was harmful to the bacteria. He had discovered penicillin.

HOW TO FIND TRUFFLES

Truffles live underground. In America they are most often found in the forests of Oregon, Washington, and the Pacific Northwest. Although trained dogs are commonly used to locate this culinary delight, if you wish to find them yourself look close to pines, firs, oaks, hazels, hickories, birches, and beeches: truffles are formed by fungi that are partners with these trees. Look for pits dug nearby, because squirrels and chipmunks love truffles.

A WITCHES' CIRCLE

When a spore falls onto the right type of soil, it germinates, and the mycelium ramifies and spreads out in the shape of a circle. As the circle widens, the central portion of mycelium, which is older, dies, leaving the shape of a ring. The fruit-bearing bodies, the mushrooms, give it away since they arise from the living mycelium by forming what traditionally has been known as a "witches' circle."

PARASITIC FUNGI

It's hard to think of fungi without associating them with parasitism, since almost all living creatures on the planet can be parasitized by some species of fungus. Any activity based on the cultivation of plants or on raising animals has to deal with problems that parasitic fungi can create. If you have ever owned an aquarium, you know that parasites are its worst enemies.

PARASITES OF CULTIVATED PLANTS

Fungi cause many serious plants diseases and may end up destroying crops entirely. In general, plants become infected after the germination tubes of the hyphae penetrate through the stomata of the leaves or through wounds in the trunk or the stalk.

CANKER, LEPROSY

Peach-tree leprosy occurs in the form of bruises on the leaves; they become twisted, pucker up, and eventually fall off; as for the fruits, they fail to develop and fall off the tree. **Canker** is a disease produced by a fungus that penetrates through wounds in the trunk and branches, producing ulcers that can even penetrate to the central core.

Root rot kills many plants without pity. The bark of the root comes off easily, and white spots joined by the fungus's mycelium appear.

Pear and apple tree scab first appear on the leaves in the form of small dark blotches. Then the fruit becomes deformed and the attacked areas shrivel and crack. The drawing shows a highly magnified view of a scab.

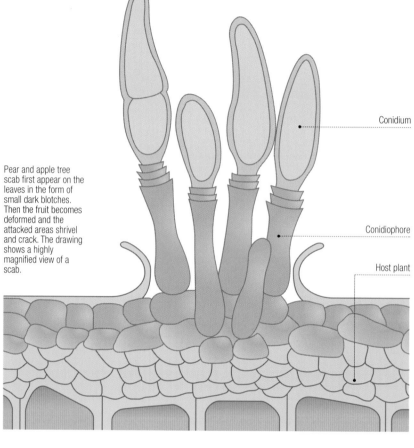

Conidium

Conidiophore

Host plant

Head smut on corn appears with tumors the size of a fist that are filled with a black powder, the spores of the fungus.

Ergot of rye. If it's in the flour used for making bread, it can cause serious psychic problems.

ANIMAL PARASITES

There are fungi that produce superficial infections where only the skin, the hair, or the nails of animals and humans are infected; examples are ringworm and athlete's foot in humans. But there are others that invade internal organs and cause more serious diseases. One of these fungi infects the housefly. Its mycelium grows inside the insect's body and uses its proteins. The insect dies in a week.

Sometimes the **specialization of parasitic fungi** gets to a point where the parasite, such as one that affects a species of beetle, can grow only on the creature's feet.

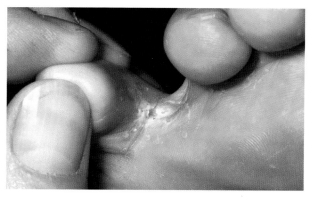

Athlete's foot is a skin disease that mainly affects the toes; it is caused by a fungus.

FUNGI THAT HUNT

It seems impossible for a plant to act as a predatory animal. But certain imperfect fungi "hunt" for **nematodes**, a type of earthworm that causes havoc on the roots of plants. The fungus acts with its mycelium. When one of these worms comes by, it forms one or more rings to catch the victim. Then it inserts its hyphae into the prisoner's body and acts like the rest of the parasitic fungi. Other fungi set traps for very tiny invertebrates and microbes.

THE FUNGUS THAT HUNTS FOR NEMATODES

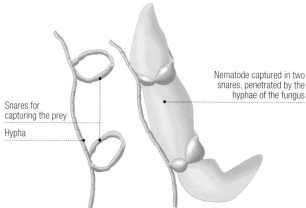

Nematode captured in two snares, penetrated by the hyphae of the fungus

Snares for capturing the prey

Hypha

From the time that plants are very small (here in a nursery), they are attacked by parasites. From this time on, fungicides need to be applied.

USEFUL PARASITES

Some parasitic fungi that attack animals and are harmful to humans, such as a plant epidemic, are used to inhibit the propagation of the harmful animal. This system of combating epidemics is **biological**; the advantage it has over chemical means is that it doesn't pollute the environment. All it takes is to cause the infection artificially by using the spores of the right parasitic fungus, or by applying them directly just as with insecticides, or else by introducing individuals previously infected in the laboratory.

Aphids or plant lice are a real plague for many fruit trees, vegetables, and decorative plants. A fungus that's the first cousin of the housefly parasite is used to combat them.

Fungi have a great calling as parasites. Don't be amazed if you see mushrooms in the woods on whose cap other, smaller mushrooms have developed that are **parasitizing their own parents**.

FUNGAL HERBICIDES

One of the major problems that farmers have always had is the weeds that compete with the cultivated plants. The way to get rid of them or control them is based on solutions that contain chemical herbicides; but parasitic fungi are now starting to be used. Of course, the herbicidal fungus has to be very specific to infect only the weeds, without hurting the crops and the environment around them, including people. The fungal herbicide is grown in laboratories and is applied just like a chemical herbicide.

FUNGAL SYMBIONTS

Many of those beautifully colored organisms you see covering bare rocks, roofs, walls, logs, and other unexpected sites are not individual beings, but close associations between a fungus and an alga that are referred to as **lichens**. This type of symbiosis also is practiced by fungi using the roots of plants, forming **mycorrhizas**, and even with animals, especially insects.

ASSOCIATING WITH ALGAE

In a **lichen**, the fungus surrounds the **alga** with its **hyphae** and sends **haustoria** into its cells. This is a **symbiosis** in which the alga uses photosynthesis to produce food for both associates and, in exchange, gets water and minerals from the fungus as protection against desiccation.

 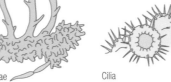 

Rhizines   Papillae   Cilia   Veins

Except for the crustaceous forms, lichens have appendices at some location on their body. If they arise on the lower surface, they help secure the thallus to the substrate.

The "mosses" on which reindeer and caribou feed in arctic regions are really lichens.

TYPES OF LICHENS

Based on how the thallus or the body of the lichen grows, there are four basic types of lichens:

- Crustaceous lichens
- Squamulous lichens
- Fruiticose lichens
- Lichens with a composite thallus

When a lichen dries, photosynthesis is interrupted, and the organism enters a period of **dormancy** that allows it to withstand very adverse conditions.

PIONEERS

Lichens can survive anywhere life is possible, since they can withstand extreme temperatures and moisture. There are lichens in equatorial jungles and northern regions where no plants can live. Along with the blue-green algae, they are the pioneers in **colonizing bare rock**: they break down the rock little by little and facilitate their disintegration by wind and rain. Thus they begin to form **soils** in which other plants can grow.

POLLUTION INDICATORS

If you see lots of lichens in some spot, you can be sure that you are breathing pure air. Lichens are the most sensitive plants to air pollution, since the toxic components in the air destroy chlorophyll. You won't find any lichens in an industrial city.

Lichens grow less than .039 inch (1 mm) in length every year. It's believed that some lichens are several thousand years old.

The reddish-purple color of Roman tunics was obtained from substances extracted from the lichens known as **urchilles**.

SUBSTANCES AND VALUABLE COLORS

Ever since Antiquity, humans have used **lichenous substances** in medicine. Today they are used in pharmaceuticals for their antibiotic, antiviral, anticarcinogenic, and anti-inflammatory properties. Lichens also have been highly sought for **natural dyes** in all kinds of colors and for quality **perfumes** because of the fresh, earthy aroma that they give off.

Introduction

Plant Anatomy

Plant Physiology

Reproduction

Flower, Fruit, and Seed

Ecology and Evolution

Algae

Fungi

Plants

Plants with Flowers and Fruits

Plants and Their Environment

Aquatic Plants

Wild Plants

Domesticated Plants

Gardens

Alphabetical Subject Index

ASSOCIATING WITH PLANTS

Fortunately, the great harm that parasitic fungi do to plants is offset by the great benefit that results from the association of other **fungal symbionts** with the **roots** of most plants. In this form of **symbiosis**, called **mycorrhiza**, the fungus benefits the plant by breaking down the organic material in the soil, making certain minerals available to the roots. For their part, the roots provide the fungus with sugars and other useful substances.

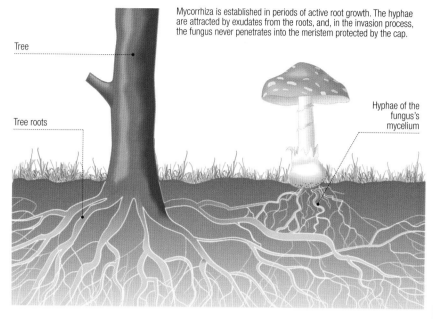

Mycorrhiza is established in periods of active root growth. The hyphae are attracted by exudates from the roots, and, in the invasion process, the fungus never penetrates into the meristem protected by the cap.

Tree

Tree roots

Hyphae of the fungus's mycelium

In **mycorrhizas**, plants joined to a single mycelium compete through their ability to capture the nutrients absorbed by the fungus and to give up to the fungus less sugars than their competitors.

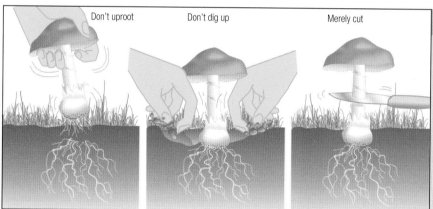

Don't uproot

Don't dig up

Merely cut

DON'T HURT THE MYCORRHIZAS

When you gather mushrooms, remember their ecological importance, since nearly all of them form mycorrhizas with the forest plants. You mustn't pull them up by the roots or dig the dirt from around them, for that would harm the mycelium of the fungus. The best practice is to cut off the carpophore, that is, the mushroom, at the base of the stalk. That way the mycelium remains intact beneath the soil, possibly spreading out and producing new carpophores.

ASSOCIATING WITH INSECTS

Fungi can even be associated with animals! The most interesting cases involve symbiosis between fungi and insects. The **leaf-cutting ants** of tropical America cultivate real "**fungus orchards**" in special chambers that they have inside their nests. In them, they accumulate ground-up leaves and excrement where the mycelium of the fungus develops. The ants feed on the special, highly nutritious hyphae without harming the rest of the fungus. And the workers take pains to keep the orchard free of "weeds."

Leaf-cutting ants hard at work

MOSSES AND LIVERWORTS

You will find mosses, those low plants that carpet the rocks, fill in cracks, or cover the rough bark of tree trunks, in damp locations. They are the simplest and most primitive land plants and are not entirely adapted to life on terra firma because they don't have real roots or seeds that can withstand drought.

THE IMPORTANCE OF THE EMBRYO

In order to liberate themselves from the aquatic environment, plants had to "invent" the **embryo**. In effect, in the life cycle of land plants, fertilized ovules or **zygotes** are kept inside the female sex organs. That way they obtain the water and nutrients from the maternal tissues that surround them and remain protected from dehydration until they begin to develop. Thus, an embryo is nothing more than a protected zygote that is not developing. In mosses and ferns, the embryo develops on the mother plant (the **gametophyte**), but, in plants with seeds, it develops away from the mother.

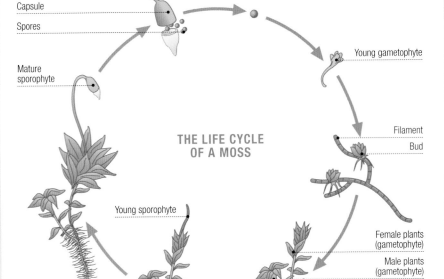

Capsule

Spores

Mature sporophyte

Young gametophyte

THE LIFE CYCLE OF A MOSS

Filament

Bud

Young sporophyte

Female plants (gametophyte)

Male plants (gametophyte)

MOSSES

Mosses are the gametophytic generation of the plant. They commonly are perennials. The sporophyte, on the other hand, is very simple, depends on the gametophyte, and always is annual and short-lived.

THE FORMATION OF A SPHAGNUM PEAT BOG

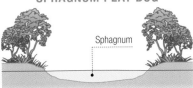

Sphagnum

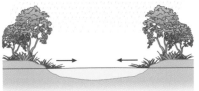

Peat

WATER FOR REPRODUCTION

Moss needs a great quantity of water or moisture to develop.

Plants that pass through the embryonic state without developing seeds in their life cycle, such as the mosses, are not entirely liberated from the aquatic environment. They need water as a vehicle so that the **male gametes** can reach the **oospheres** and fertilize them. For that reason, the mosses and liverworts live in damp locations, and fertilization is carried out with the use of rain or dew.

WHAT IS PEAT MOSS?

This type of black, spongy, light earth that is used as a substrate for potted plants is called peat moss. It's a carbonaceous material made up of partly decomposed plant remains that accumulate in saturated soils that have no oxygen. **Sphagnum** mosses and other mosses produce peat without the need for standing water: as they grow at the apex, they die at the base, which turns into peat moss because it retains rainwater for a long time.

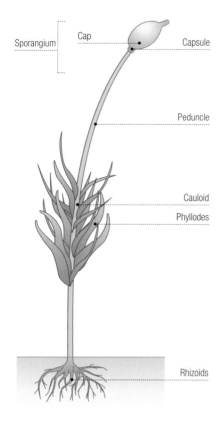

Sporangium — Cap — Capsule

Peduncle

Cauloid

Phyllodes

Rhizoids

THE ORGANS OF A MOSS PLANT

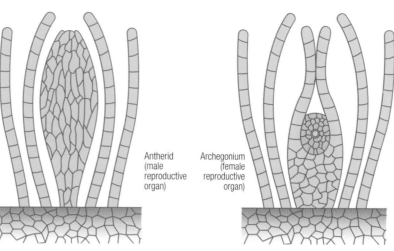

Antherid (male reproductive organ)

Archegonium (female reproductive organ)

THE BODY OF MOSSES

Mosses have neither vessels to carry sap nor support tissues. They don't have a real thallus or true leaves, and, instead of roots, they have **rhizoids** for use in attaching themselves to the substrate. But they can absorb water and nutrients through any cell of their body or **thallus**. Mosses never grow more than a hand high. What is commonly called the "fruit" of the moss is the sporophyte, which consists of a peduncle (**mushroom**); at the end of this peduncle, there forms a sporangium provided with an opening through which the plant releases spores when there's an opportunity for the wind to disperse them.

LIVERWORTS

These plants are even simpler than the mosses. The body commonly is a flattened thallus or a type of stem covered by two lateral rows of phyllodes without veins. Inside the sporangia, they have special cells known as **elaters**, which facilitate the dispersion of the spores. Many liverworts are aquatic and live near springs. Others are epiphytes that live on trunks, branches, and leaves in tropical rain forests.

LIVING SPONGES

Perhaps you have noticed that sellers of decorative plants sometimes use mosses in packing the roots of certain plants. These mosses are capable of absorbing and retaining large quantities of water. They are ideal for keeping the roots moist until the plant is put into the ground.

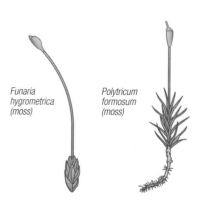

 Along with lichens, mosses share the roles of pioneer in colonizing bare areas and of pollution indicators.

 Mosses grow even in dry, arid areas, but they spend most of their life in the form of spores. When it rains, they develop and die in a short time.

 There are mosses that can store up to twenty times their dry weight in water.

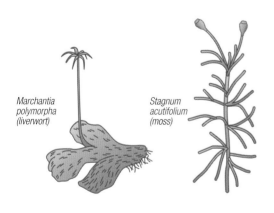

Marchantia polymorpha (liverwort)

Stagnum acutifolium (moss)

Plagiothecium undulatum (moss)

Funaria hygrometrica (moss)

Polytricum formosum (moss)

FERNS, LYCOPODS, AND HORSETAILS

Ferns are very easy to distinguish from other plants because of the peculiar leaves that they have, known as **fronds**. They often are used as decorative plants inside houses, but they also are found in nature near springs and other damp areas. Some of them resemble certain mosses and liverworts. However, there is a great difference between these two groups of plants: the leaves of ferns have **conducting vessels**, and those of mosses don't. Ferns are **vascular plants**; mosses are not.

CONDUCTING VESSELS AND LIGNIN

It can be said that one of the most important steps that plants took along their evolutionary history was the development of conducting vessels (**phloem** and **xylem**) and the ability to synthesize **lignin**, the substance that gives plants rigidity, making it possible for them to stay erect on terra firma and be protected from herbivores. The first plants that took this step were the ferns.

FORESTS OF FERNS

It may strike you as strange that the ferns that you commonly see are so short. There also are **arborescent ferns** but in very few areas around the world, because most of those ferns disappeared millions of years ago when the modern trees appeared, which are better adapted to current environmental conditions. Before the modern trees appeared, most of the planet's forests were composed of ferns, lycopods, and horsetails.

FERNS AND MOISTURE

Why do we find ferns only in damp locations? The reasons are the same as for the mosses: in order to reproduce, they still need water. The **antherozoids** that are liberated from the **antherids** after a rain have flagella, and they swim toward the **archegonia** and fertilize the **ovule**, thereby producing a **zygote**.

The spores liberated by the sporangia located on the fronds germinate and give rise to a gametophyte that produces male and female gametangia (antherids and archegonia, respectively) when it matures.

The plant that we call a fern is the sporophyte. The gametophyte, which is known as a prothalium, is very small, and we usually don't see it.

The zygote that results from the fertilization of the ovule gives rise to an embryo that becomes independent of the gametophyte when it puts out its first roots and leaves.

A large part of the **bituminous coal** that humans have mined from underground originated in the dead fern trees, lycopods, and horsetails that lived many millions of years ago.

FERN FRONDS

Veins

What makes it possible to tell if a plant is a fern is its leaves or fronds, which unfurl as they grow. In general, they are composite leaves with clustered masses of **sporangia**, known as **sori**, on the underside. They also have the **veins** that characterize vascular plants.

Sori

Rachis

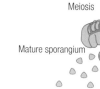

THE LIFE CYCLE OF A FERN

Adult sporophyte

Sporangia

Meiosis

Mature sporangium

Young sporophyte

Fertilization

Mature gametophyte

Spore

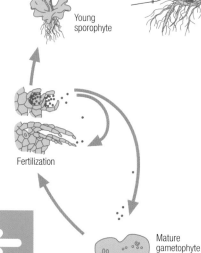

LYCOPODS

The lycopods or club ferns are part of the fern family, but they exhibit some noteworthy differences. They consist of a creeping stem that sends up erect thalluses covered with tiny, flat, thin leaves arranged in a spiral known as **microphylls**. At the ends of these thalluses are the **sporophylls** or specialized leaves arranged like pinecones, on which the **sporangia** form.

SPORES AND FIREWORKS

If you approach a lycopod with the **sporangia** opened, you merely need to shake the plant gently over a paper and you will have what's known as **plant sulfur**. These are the yellowish **spores** of the lycopod. If you toss them into a flame they produce a spectacular flash because they are very flammable. Plant sulfur has often been used in making fireworks.

TYPICAL LYCOPOD

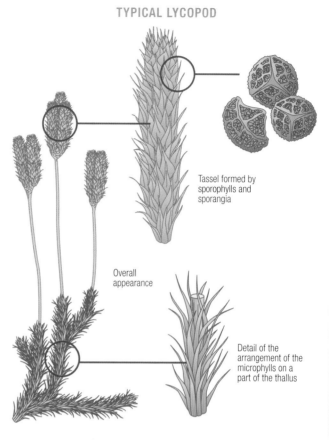

Tassel formed by sporophylls and sporangia

Overall appearance

Detail of the arrangement of the microphylls on a part of the thallus

HORSETAILS

Equisetums, or horsetails, also are very close relatives of the ferns. The **sporophyte** of the horsetails is made up of a horizontal, ramified, underground **rhizome** from which the jointed, aerial thalluses grow that give this plant its name. Rings of small branches with little scale-shaped leaves grow from the joints of these stems. The spores are contained inside the **strobili** that appear at the end of certain branches.

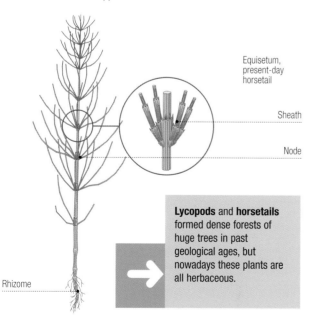

Equisetum, present-day horsetail

Sheath

Node

Rhizome

Lycopods and **horsetails** formed dense forests of huge trees in past geological ages, but nowadays these plants are all herbaceous.

HORSETAIL SCRUBBERS

Before the modern aluminum and steel wool dish scrubbers were invented, horsetails were used for scrubbing pots and pans, as well as for polishing metals. Their abrasive nature is caused by the **silica** deposits present in the **epidermis** of these plants.

Ferns generally are found in moist areas, mixed in with other plants and trees.

Introduction

Plant Anatomy

Plant Physiology

Reproduction

Flower, Fruit, and Seed

Ecology and Evolution

Algae

Fungi

Plants

Plants with Flowers and Fruits

Plants and Their Environment

Aquatic Plants

Wild Plants

Domesticated Plants

Gardens

Alphabetical Subject Index

PLANTS WITH NAKED SEEDS

Most of the plants that you commonly see are seed-producing plants, since they are the ones that dominate the planet's exposed land. That's because they can carry out fertilization in the air by means of **pollination**, which makes the presence of surface water unnecessary for reproduction. However, there are differences between the reproductive system of a pine and that of a cherry tree. The former has naked seeds; a cherry tree has them encased in a fruit, the cherry.

CONIFERS

Plants with naked seeds are known as **gymnosperms**, but nowadays the great majority of these plants are **conifers**, a name that they owe to their principal characteristic: the cones in which their reproductive organs are located. Normally, conifers have male and female cones on the same individual; in other words, they are **monoecious**. They all are woody (trees or shrubs), and, for the most part, they keep their leaves year-round. They don't produce flowers, and their seeds are simply inserted into woody pinecones.

A MALE PINECONE

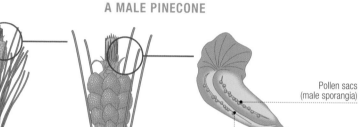

Pollen sacs (male sporangia)

Pollen grain

 A pollen grain from a conifer can take more than a year to form a pollen tube, so there can be a long time between pollination and fertilization.

THE "INVENTION" OF POLLINATION

The union of the male and female **gametes** through pollination was a true revolution in the plant world, and it became possible only with the development of the **pollen tube**. When the pollen passes between the **scales** of a **female cone**, they close and the **pollen grain** starts to **germinate**; it lengthens, sending out a tube that reaches the oosphere, where the sperm nuclei or antherozoids are released. One of them fertilizes the **oosphere** and the **zygote** is formed, which turns into a **seed** equipped with wings and nutrients, ready to be carried off by the wind.

THE PARTS OF A PINECONE OR A FEMALE CONE

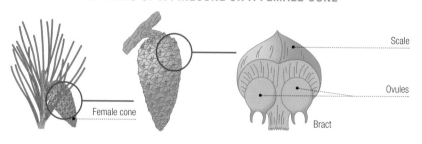

Scale

Ovules

Female cone

Bract

THE LIFE CYCLE OF A CONIFER SUCH AS A PINETREE

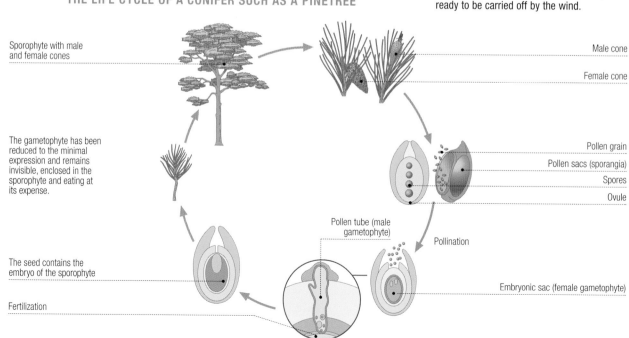

Sporophyte with male and female cones

The gametophyte has been reduced to the minimal expression and remains invisible, enclosed in the sporophyte and eating at its expense.

The seed contains the embryo of the sporophyte

Fertilization

Male cone

Female cone

Pollen grain

Pollen sacs (sporangia)

Spores

Ovule

Pollen tube (male gametophyte)

Pollination

Embryonic sac (female gametophyte)

A SPECIAL LEAF

Only a few conifers are deciduous. Most are evergreens and have foliage that is well adapted to withstand hot, dry summers and cold winters. This adaptation consists of having many but very tiny, coriaceous leaves protected by a thick cuticle. They can be in the shape of a needle or a scale, or else can be flat.

VARIOUS TYPES OF CONIFER LEAVES

Pine acicles (or needles)

Squamiform leaves of the cypress

Flattened leaves of the yew

SOME FAMILIES OF CONIFERS

Family	Species
Pinaceae	Pine, spruce, hemlock, cedar, larch
Araucariaceae	Araucaria
Taxodiaceae	Sequoya, swamp cypress
Cupressaceae	Cypress, thuya, common juniper, Savin juniper
Podocarpaceae	Podocarp
Taxaceae	yew

All the organs of the yew are poisonous, with the exception of the **fleshy red cupule** that covers the seed. Even the seed inside the cupule is poisonous.

FAMILIES OF CONIFERS

Even though the conifers have lots of things in common, botanists group them into a series of families. Some of these families are not represented on all continents; but it's easy to get to know them because all conifers are very decorative and are present in parks and gardens all around the world.

SURVIVORS FROM ANCIENT TIMES

Cycads and **ginkgos** are plants that frequently are used in cities because they are very decorative. They too are plants that have naked seeds, but they are not conifers; they belong to plant families that abounded millions of years ago and now are nearly extinct. Cycads resemble palm trees, but they don't grow as high. The leaves of the ginkgo are fan-shaped, and, in the fall, they turn yellow before falling.

The elegant cypress has a very straight trunk and an elongated, spindle-shaped crown. It commonly is used to decorate gardens and cemeteries.

The Douglas fir, which is found mainly all over North America, can reach heights of 280 feet (90 m).

The Savin juniper is a shrubby plant that can reach 33 feet (10 m) and live hundreds of years.

DICOTYLEDONS

There have not always been plants with flowers growing on the face of the earth. Flowering plants, called **angiosperms**, appeared in the days of the dinosaurs in an attempt to adapt better to land conditions. In contrast to the gymnosperms that bear naked seeds, the angiosperms enclose their seeds inside a **fruit**. In many of them, the **embryo** contained in the seed has just one leaf of **cotyledon**, but there are more that have an embryo with **two cotyledons**, which are, therefore, known as **dicotyledons**.

NEW ENERGY-SAVING INVENTIONS

The female cone of a conifer is an investment that the plant makes whether or not fertilization takes place. An angiosperm, on the other hand, doesn't start to invest much energy until fertilization is guaranteed. Only when fertilization has taken place, does the seed mature (and the flower drops off), forming the embryo and the necessary nutrients for the first steps of its development.

Then the ripe seed can remain inside a broad variety of fruits.

PERFECT FLOWERS

Dicotyledons usually have flowers with male and female reproductive organs (stamens and pistils, respectively)—that is, **perfect flowers**, although there are exceptions. This property facilitates pollination and self-fertilization.

DECIDUOUS PLANTS

These are the plants that lose their leaves in the fall and sprout new ones in the spring. Their leaves are flat, tender, relatively large, and have a very fine cuticle. Like all dicotyledonous plants, they grow not only in height, but also in girth, like the conifers.

THE LIFE CYCLE OF A DICOTYLEDON SUCH AS THE APPLE

Pollen grain
Stigma
Stile
Pollen tube
Pollen
Ovule
Ovary
Embryonic sac
Pistil

Pollen sacs
Stamen — Anther
Filament
Petal
Pistil
Sepal
Ovules
Receptacle — **Perfect flower**

Developed ovary

Ovary
Fruit

Adult sporophyte

Seed

Seed germination

Young sporophyte

Almost all woody plants (trees and shrubs) are dicotyledons.

The great advantage that plants with flowers enjoy is that they are capable of growing and generating seeds quickly. Conifers take longer.

The beech is a good example of a deciduous tree. In the fall, its leaves turn yellow and drop off. The beech makes very beautiful woodlots.

DECIDUOUS BUT EVERGREEN

Where the mildness of the climate permits continued biological activity, through favorable temperatures as well as humidity, the leafy plants hang onto their leaves throughout the year, as in the tropical jungles.

Latex is obtained by tapping the rubber tree or hevea; tires for vehicles were once made by vulcanizing the rubber.

Here is an olive tree that's nearly ripe for the harvest; the olives can be eaten as a garnish or used as a source of the highly valued olive oil.

EVERGREENS WITH CORIACEOUS LEAVES

In Mediterranean-type climates, dicotyledons, such as the holm oak and the shrubs that accompany it, as well as a great number of aromatic plants, also remain green year-round, but they have small, leathery leaves. Just as with the leafy evergreens, they lose their leaves throughout the entire year but on a reduced scale and constantly replace them.

VEGETABLE PLANTS AND FRUIT TREES

The vast majority of vegetable plants, fruit trees, and wild fruits are dicotyledons. Some of them, such as peas and beans, germinate quickly under favorable conditions and make it easy to see the two cotyledons that characterize the embryo of all dicots.

THE GREAT OPPORTUNITY

It's probable that flowering plants began to gain ground on the other plants because of the **migratory dinosaurs** that tore up and fertilized the ground over which they passed. No other group of plants could compete with the flowering plants in speed of producing seeds and colonizing ground.

Grapes come from vines; this fruit can be eaten fresh, and when fermented, it produces wine.

CACTI AND FLESHY PLANTS

The perfect flowers of cacti often are very beautiful, and they reveal that the cacti are dicotyledons. One way to recognize a dicotyledon by its flowers involves counting the number of pieces in each part of the flower (petals, stamens, pistils). Almost all dicotyledons have four or five pieces or several groups of four or five.

ATTRACTIVE FLOWERS AND LEGUMINOUS PLANTS OF THE PRAIRIES

Almost all grasses and plants with attractive flowers, such as the carnation and the rose, as well as all legumes such as clover and alfalfa are dicotyledons. In general, plants that have ramified veins in their leaves tend to be dicotyledons.

Cacti are thick plants that produce an edible fruit known as the prickly pear.

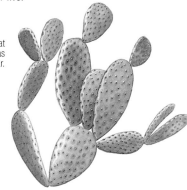

The daisy is an herbaceous plant from the family of composite plants, which come in many species and colors.

MONOCOTYLEDONS

Plants that have flowers whose embryo has just one leaf or **cotyledon** are known as **monocotyledons**. But you shouldn't place too much emphasis on this characteristic, because there are other more important ones. You merely have to compare a palm tree to an oak, which increases the girth of its trunk and branches year after year, while the palm tree gains only in height. Now that's a noteworthy difference!

THE DIFFERENCE BETWEEN MONOCOTYLEDONS AND DICOTYLEDONS

In addition to differing in their embryonic structure, both groups of plants with flowers and fruits exhibit other important differences.

DISTINCTIVE FEATURES OF MONOCOTYLEDONS AND DICOTYLEDONS

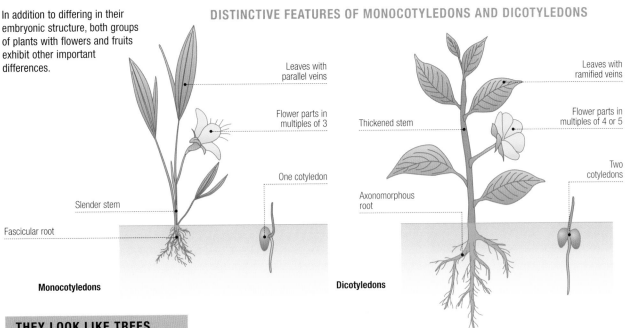

Leaves with parallel veins

Flower parts in multiples of 3

One cotyledon

Slender stem

Fascicular root

Monocotyledons

Leaves with ramified veins

Thickened stem

Flower parts in multiples of 4 or 5

Two cotyledons

Axonomorphous root

Dicotyledons

THEY LOOK LIKE TREES

A banana tree is a tropical plant that produces bunches of fruit (bananas) at the end of its axis of inflorescence. It looks like a tree, but what you would call the trunk is really nothing more than a shaft surrounded by the sheaths of the leaves that grow from a short rhizome located at the base. So, in reality, this is a monocotyledonous "large grass" instead of a dicotyledonous woody tree.

A COMPARISON OF SOME CHARACTERISTICS

Monocotyledons	Dicotyledons
Embryo with one cotyledon	Embryo with two cotyledons
Ripe seeds with endosperm	Ripe seeds with no endosperm
Leaves with parallel veins and smooth edges	Leaves with ramified veins
No growth in thickness	Growth in thickness
Flower parts in threes or multiples of three	Flower parts in fours or fives, or multiples of these numbers
Generally herbaceous	Herbaceous and woody
Vascular bundles dispersed in thallus	Vascular bundles in thallus form a cylinder
Fascicular root (branching with equal development)	Axonomorphous root (one primary root plus secondary roots)

THE GRAMINEAE

These are the most abundant grasses that make up natural grassy areas and are planted by humans and are suited for pasturing animals, since the thallus and the buds remain at ground level. Their flowers are not particularly attractive, but the seeds of the members of the cultivated **cereals** family have been very important for humans from time immemorial. **Bamboo** is the only gramineous plant that is not herbaceous.

Rice is a basic food for much of humankind. The photo shows terraced rice paddies in Bali, Indonesia.

IRISES, AGAVES, ONIONS . . .

These plants are monocotyledons that are adapted to survival in dry regions by means of **bulbs**, **tubers**, or **rhizomes**. They include some very well-known flowers, such as the lily, the iris, the tulip, and the gladiolus. Others have a coriaceous thallus, such as asparagus, the dragon tree, and the yucca. Onions and garlic are representative of the minority of produce that is monocotyledonous.

The yucca is an arborescent plant that reaches 50 to 60 feet (15–20 m) in height; when it blooms, it produces large, hanging flowers.

The dragon tree is an arboreal plant that grows very slowly, but it can grow to large sizes over the thousands of years that it lives.

ARUMS AND PALMS

These plants have in common the type of **inflorescence** that their flowers form, which is known as a **spadix**. Otherwise, they are very different from one another. Palm trees can't withstand cold climates and are among the few monocotyledons that look like trees, even though their stem grows only in height and always maintains the same diameter.

The coconut tree is a palm of which all parts can be used: the trunk is used in carpentry; the terminal bud is a fine food; the leaves are used in roofing huts and making baskets; the milk inside the fruit, the coconut, is used as a drink, and the meat can be eaten fresh or dried, and an oil can be pressed from it and used to make soaps and cosmetics.

The dwarf fan palm or palmetto is the only living European palm tree. It doesn't grow as tall as the other palms. The heart of the trunk is edible and is known as *heart of palm*.

ORCHIDS

There also are orchids in temperate regions, but they are vigorous herbaceous plants with less spectacular flowers than those of tropical orchids. The originality of these plants is because of an adaptation to pollination by insects. The flowers are very irregular, and there is almost always one petal that is larger than the others and forms a type of lip.

There are more than 20,000 different types of orchids in the world; some of them are grown in greenhouses and sold for very high prices.

REEDS, PAPYRI, AND BULLRUSHES

These plants live in swampy terrain or other areas where standing water accumulates. They are plants that have **rhizomes**, and, as a result, they are perennials. In antiquity, sheets were made from the pith of the papyrus and were used as writing paper. The stalks of these plants can reach ten feet (3 m) in height.

Papyrus is a robust plant with a stem that can grow to ten feet (3 m) tall and four inches (10 cm) thick. Formerly, these stems were used as a source of strips that were arranged perpendicular to one another, moistened, and pounded flat with a mallet to produce a writing surface.

HORCHATA OF CHUFA

The chufa is a species of reed that has been cultivated for centuries in the eastern Mediterranean. It is a perennial plant whose rhizome puts out underground stolons that form small ovoid tubers known as **tiger nuts**. Tiger nuts are used to make a popular drink called *horchata* in Spain and Mexico. In the United States you may buy horchata-based drinks and snacks, which have a distinctive sweet, chestnut-like flavor, at many Latin markets and some health food stores.

COLD REGION PLANTS

The cold regions of our planet are the places farthest from the equator and closest to the poles. The plants that live in these places are highly conditioned for the annual low and high temperatures and the length of the yearly period in which they can be biologically active. The two basic types of biome that are found in these areas are the **tundra**, in the coldest area, and the vast **conifer forests** that make up the **taiga**.

THE ABODE OF THE REINDEER AND CARIBOU

The reindeer and caribou, inhabitants of the **tundra**, have to live by continually migrating from one place to another because there isn't enough vegetation in any local territory to support them. The soil of the tundra is like a frozen swamp. A superficial layer of the soil no deeper than a hand's breadth thaws only during the extremely short summer. In these conditions, the only plants that can live are **lichens**, **mosses**, **reeds**, and some **very low shrubs**.

Some areas in the south of the Argentine and Chilean Patagonia have vegetation of the steppes and even the tundra.

Map of the Arctic, with the Arctic Circle, showing the limits of the tundra.

The word **tundra** comes from a Finnish word that means "land without trees."

DWARF TREES

In the more temperate tundra, there are also **alders**, **birches**, and shrubby **willows** that grow in addition to the usual plants; they are first cousins of the large trees that we know by the same names. They are practical examples of the different routes by which life evolves by adapting to different environmental conditions.

The leaves of the dwarf birch (left) and the dwarf willow (right), which can resist the rigors of the tundra.

IGLOO PLANTS

It's possible to live inside the igloos that the Eskimos construct from blocks of ice even though the outside temperature is 60°F below (−50°C), because ice is a very effective thermal insulator. Many plants of the tundra and the high mountains do the same: they adopt a cushion form that acts like an igloo when it becomes covered with snow.

In order to survive low temperatures, the saxifrage pad adopts the form of an igloo.

THE DOMINATION OF THE CONIFERS

There are conifers all over the world, but the place where they reign and no other plant can compete with them is in the cold areas where conditions are less severe than on the tundra.

There they form extensive, dark evergreen forests of **spruces**, **hemlocks**, **pitch pines**, and other conifers.

In the understory, there are blueberries, mosses, lichens, and lycopods. On the North American continent, the species are different: giant **sequoias**, **firs**, and **Douglas spruce**.

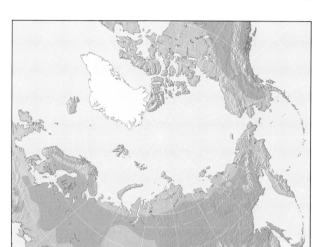

Map of the Arctic, with the Arctic Circle, showing the limits of the taiga.

Some species of spruces (like this Douglas spruce in the western United States) reach over 300 feet (100 m) in height and a trunk diameter of up to 16 feet (5 m).

The tundra is covered with snow from 200 to 300 days a year.

In the border between the taiga and the tundra, the average daily temperatures are below freezing for eight consecutive months.

FORTY BELOW ZERO!

How can **spruce** needles withstand temperatures down to 40°F below without being damaged by ice? Because they have the least possible surface area, are coriaceous, and are covered with a thick layer of insulating wax. As if that weren't enough, when winter arrives, these needles dehydrate, so they can't freeze, and the tree sinks into a deep **dormancy** until decent weather returns.

A DIFFERENT CONIFER

One of the few conifers that lose their needles is the **larch**. This seems to contradict everything we have said about the advantages of the conifers, but that's not the case. Even though the needles of the conifers withstand very low temperatures, very few of them can survive the rugged Siberian winter. The larch has adapted by dropping its needles every year. That way it can live in places where the spruces can't.

Although the larch is a conifer, it loses its needles (in the illustration) during the severe winter.

ALLIES IN POOR SOIL

The soil in conifer forests is neither deep nor rich in **mineral nutrients**. The fallen conifer needles form a low-quality **humus** that mineralizes very slowly, so the nutrients are found in the superficial layer of the soil. Fortunately, there are many **fungi** that live in these forests; they form **mycorrhizas** with the conifer roots, absorbing nutrients that they then yield to the tree.

THE GREAT ADVANTAGE

It's not resistance to the cold that favors the conifers in the taiga, because leafy trees also get through the winter well by dropping their leaves; instead, it's the ability to keep their leaves all year long. After a long winter, the conifers start to photosynthesize as early as the first day of spring, and they also save energy by not having to produce new leaves every year. Considering that the favorable season on the taiga is very short, this is an indisputable advantage for the conifers.

Canada has the largest conifer forests in the world.

DECIDUOUS FORESTS

Whenever you see woodlands of beech trees, oaks, maples, chestnuts, and other **deciduous leafy trees** you can be sure that you are in a region with a temperate and moist climate with clear delineations between spring, summer, fall, and winter and rain throughout the year. These forests, which are referred to as **deciduous** (seasonally shedding), are the best adapted to these conditions, as long as they have a soil that's rich in nutrients, since replacing the foliage every year requires quite an expenditure of energy.

DELICATE LEAVES THAT SQUANDER WATER

The leaves of luxuriant plants are tender and flat, and they are covered only with a very fine cuticle. This makes them very efficient at moderate temperatures and in the presence of plenty of water, but it would be a major problem in the face of a freeze or a hot, dry summer. In a freeze, the water contained in the tissues of these leaves would form ice crystals that would destroy them. Also, leafy plants transpire such a quantity of water through their leaves that they would become dehydrated under the sun and drought of a Phoenix summer.

In the fall, it's very easy to distinguish the evergreens (green) from the deciduous trees, which turn colors.

The almond is a fruit-bearing tree that's very important from an economic viewpoint, since almonds have many uses in foods; however, its precocious flowering makes it very vulnerable to late cold spells.

RAIN IS NOT CRITICAL

You can find deciduous forests in places where there's not much rain for the normal requirements of these plants. The important thing for the plants isn't so much the amount of water that falls, but instead, the difference between that amount and what evaporates, either directly from the soil or through the leaves (**transpiration**). So a forest of leafy plants can exist in an area that doesn't get much rain if there are frequent fogs or clouds that cut down on evaporation.

WHY SHED THE LEAVES?

You may think that a leafy plant drops its leaves before going into winter to keep the ice from destroying them. There's another reason we have to add to this one, and it's at least as powerful. Frozen water can't be absorbed by the roots, so in the winter, when the soil is frozen, a beech would die of thirst no matter how little its leaves transpired. It's better for it to get rid of the leaves and go into a period of dormancy until the arrival of spring.

The elm (here, an English elm) is a riparian tree that likes cool, moist areas.

Birch wood is white and lightweight, and good for turning on a lathe.

A normal beech grove transpires some 4,000 tons of water every year.

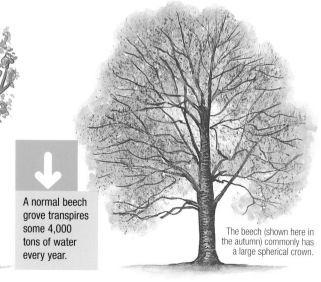

The beech (shown here in the autumn) commonly has a large spherical crown.

THE SEASONS AND THE CYCLE OF THE FOREST

Introduction

Plant Anatomy

Plant Physiology

Reproduction

Flower, Fruit, and Seed

Ecology and Evolution

Algae

Fungi

Plants

Plants with Flowers and Fruits

Plants and Their Environment

Aquatic Plants

Wild Plants

Domesticated Plants

Gardens

Alphabetical Subject Index

THE WINTER REST

The trees in a deciduous forest take a rest after dropping their leaves in the fall. Their dormant buds, sheltered from the cold under a protective cover and dehydrated to withstand freezes, can survive temperatures below −15°F (−25°C).

 The leafy tree that beats all the records in resisting low temperatures is the birch. Its hibernating buds can resist temperatures down to 40°F below.

SPRING

When spring arrives, the leafy plants begin intense photosynthetic activity and develop large crowns. This is the time that the plants in the understory take advantage of to prosper, before the trees close off the sky and plunge the forest into shadow.

SUMMER

Now the sun's rays don't get through the forest, which reduces water loss from the soil. In the hottest hours of the day, the plants close their stomata and remain inactive until nightfall, when they open the stomata once again.

FALL

When fall comes, the forest trees gradually reduce the water supply to the crown and withdraw the chlorophyll from the leaves. That's how they prepare to shed the leaves. But, before they do that, they produce the buds for the new leaves that will sprout the following spring. Finally, the ground gets covered in dead leaves that produce a high-quality humus for starting a new cycle.

A LIGHTER FOREST

Leafy plants don't always form pure, shady deciduous forests. Often the beech is not the climax plant, and it lives with various species of pines, oaks, and yews. These are known as **mixed forests**, which often are inhabited by many kinds of animals, because they offer a great number of possibilities, that is, ecological niches. A mixed forest is less monotonous than a typical deciduous forest.

OAKS

When all the deciduous trees have lost all their leaves, you can see that the oak hangs onto them in partly withered form until just before the spring growth. This is a special tree that occupies an intermediate position between the leafy plants and the trees with coriaceous evergreen leaves like the oak. As a result, oak groves are found in areas where the climate is temperate.

OPPORTUNISTS

Botanists speak of opportunistic species that quickly take advantage of clearings in the forest to thrive, whether through the falling or the death of a tree or as the result of a fire, since they can't thrive in the shade of a closed forest. **Maples** are opportunistic trees. Maples don't form forests of their own. They are most commonly found scattered irregularly in a forest. In the fall, you can identify them immediately by the reds and yellows of their leaves.

Maple leaves.

THE PLANTS OF MEDITERRANEAN CLIMATES

If you look closely at an oak leaf, you will see that it's coriaceous (leathery), lustrous, and somewhat thicker than the leaf of luxuriant plants. This is the type of leaf that characterizes the plants that are adapted to a Mediterranean climate, with very mild winters and long, dry, and hot summers. These are plants that stay green year-round, even though they experience thirst and, therefore, hunger, because without water plants can't carry out photosynthesis.

LEAVES THAT SAVE WATER

Mediterranean plants need to save water in order to survive. So they have small leaves that present little surface area for evaporation and which are covered by a thick cuticle of waterproof wax, that way they can control evaporation through the stomata, which are located only on the underside of the leaves.

The leaves of the holm oak are three or four years old; after that time, they fall off without turning yellow, around the month of August, when the tree is in vegetative repose, so it can withstand the extreme summer heat. The fruit of the holm oak is the acorn.

 A wild olive tree can produce several million flowers, but only a very small portion of them will grow into ripe fruit.

The olive is a typically Mediterranean tree with a great lifespan, since some specimens live for thousands of years.

A COSTLY LEAF

If you consider that an oak "is starving" and that its robust leaf has a higher production cost that the fine leaf of a luxuriant plant, you will understand that it's more economical for the plant to hang on to the leaf instead of dropping it every year and having to grow it all over again.

Cork oak meadows are the most valuable of all. In addition to offering the same advantages as other meadows, they also produce **cork**.

TAKING ADVANTAGE OF WINTER

The Mediterranean fall and winter are not cold, so plants can continue their photosynthetic activity—feeding themselves, in other words—on most, if not all, days. In addition, this is precisely the time of year when the plant has the most water for nourishment. So it's no wonder that Mediterranean plants keep their leaves all year long. That way they can make up for the thirst and the "hunger" that they have had to endure throughout the prolonged summer.

HOLM OAK AND WILD OLIVE MEADOWS

The meadows of the Iberian Peninsula are an example of the wise use of nature's products without harming it. There are meadows of holm oaks, wild olives, and mixtures of both species. The livestock grazes at will in a meadow, feeding on roots, tubers, grass, tender branches, and fruits (acorns and wild olives), and the animals fertilize the earth with their droppings. In addition, the crowns of the trees protect the animals from cold and heat.

After the olive tree, the carob tree is the Mediterranean evergreen tree most often cultivated by humans.

REDUCED GROWTH

Another characteristic of Mediterranean vegetation is the reduced size of the trees in comparison with the huge trees of the leafy temperate forests. In the most arid regions, the dominant vegetation consists of shrubs that have deeper roots than even the trees.

The tree strawberry is a bushy plant that reaches 12 feet (4 m) in height. Its fruit is granular, red in color, and edible.

MEDITERRANEAN PINE FORESTS

In certain very arid areas, the pines get ahead of the holm oaks. But most of the current Mediterranean pine forests are the result of cutting holm oaks.

AN EXCEPTIONAL PINE

The pines that you commonly see are not resistant to fire; their buds can't survive such high temperatures. But there is a pine that grows on the Canary Islands that produces new buds after a fire. It's the **Canary pine**, a huge tree that can grow up to 185 feet (60 m) tall, with a trunk 8 feet (2.5 m) in diameter.

LAUREL WOODS

This is a temperate, evergreen forest that develops in a **benign climate** with very little contrast between seasons and with enough moisture to sustain biological activity indefinitely. This type of ecosystem is typical of the Mediterranean region. The laurel woods closest to the Iberian Peninsula are located in the Canary Islands, where much of the moisture comes not from precipitation but from the condensation of fog and clouds.

Needles of the Canary pine (on the island of La Palma)

Leaves, flowers, and fruit of the laurel

THE MAQUIS

You mustn't think it's an easy task to make a trail through the thick, shrubby vegetation typical of the Mediterranean known as the maquis. Unfortunately, it will be hard for you to experience it in its full splendor, since fires and urban development have ravaged these plant formations. But in certain protected natural areas you can still admire the beauty, the colors, and the perfume of the maquis.

DESERT PLANTS

The deserts are the areas of the planet where it almost never rains. But there are plants that are adapted to living under minimal conditions, as we all have seen in countless western movies shot in the deserts of Arizona, Sonora, and Baja California. One characteristic of the desert landscape is vast, empty spaces between plants, as well as an absence of trees.

MANY TYPES OF DESERT

In spite of the aridity that characterizes deserts, every one of them has its own personality. There are dry, hot deserts like the Sahara; very arid but cool ones, like the Atacama of Peru and Chile; and dry and very cold ones, like the Puna in the high plateau of the Andes, which is located at an altitude of over 12,000 feet (4,000 m).

The trunk of the **saguaro cactus** has folds like an accordion; it swells when the cactus stores water and shrinks as the water is used up.

The saguaro defends itself with its thorns against thirsty animals that would like to gnaw its juicy trunk. When saguaros begin life, they can't survive in direct exposure to the sun; that's why they always originate clustered at the foot of bushes. After their first stages of development, they outstrip their "nursemaids."

ROOTS FOR ARIDITY

Not all desert plants use the same strategy to cope with lack of water. The gigantic **saguaro cactus** from the Sonora Desert has a long root network that sometimes extends for more than 100 feet (30 m); however, these roots are quite shallow, because rain is scarce but torrential, and it only waters the upper layer of the soil.

Desert plants have a thick epidermis to keep the water they hold inside from evaporating.

Some deserts on the planet are totally devoid of vegetation.

PROTECTED WATER

The desert soil needs to be totally different from the soil of the humid regions. Whereas the soil of a wet climate is composed of fine grains, the desert terrain is characterized by thick, grainy sand that allows water to penetrate fast and deep, and thus to be unaffected by the rapid evaporation taking place in the upper layer.

DISPOSABLE LEAVES

There are plants that deal with the aridity of a desert in a similar fashion to that of the Mediterranean plants by having very small, leathery, tough leaves. But others have broad leaves without much protection, and they don't even bother to close their stomata at the most critical times; when drought pinches, they drop their leaves, regardless of the season. But, while they have their leaves, they have twice the photosynthetic capacity of other plants.

Tillandsias are specialists in capturing water from fog. Their leaves have absorbent hairs that directly capture the droplets of condensed water from fog.

Opuntias are a genus of fleshy plants typical of arid desert areas.

The **jojoba** plant is the source of a liquid wax widely used in the pharmaceuticals and cosmetics industry.

THE DECEPTIVE DUNE STABILIZER

If you saw a **mesquite** you would think you were looking at a low plant. However, it's really a tree. From the time of the mesquite's birth, sand accumulates all around it through the action of the wind. The mesquite produces new branches that emerge from the sand, and, with time, a dune forms that is held together by a powerful tree several yards tall. Only the branches of the crown stick out from the sand.

OASES

Oases are islands of vegetation in the middle of the desert. They are always in depressions (hollows) beneath which water accumulates through filtration. The water is not visible from the surface, but the roots of the plants owe their life to this water, which sometimes is very close to the surface. In contrast to normal desert plants, oasis plants commonly have deeper but less widespread roots, since they are tapping into water deep underground.

THE TOUGHEST ONES

The plants best adapted to extreme dryness are the ones that absorb great quantities of water when it rains and store it in large cells that they have in their leaves or stems (cacti). They are known as **succulent plants**. **Cacti** have no leaves or have leaves that have been transformed into thorns.

The water contained in the leaves, stems, and roots of certain **succulent plants** of the desert are used by inhabitants and travelers in emergency situations.

THE PLANTS OF TROPICAL JUNGLES

In the equatorial regions that enjoy abundant rainfall, there are two factors that come together and tremendously favor plant development: warmth and moisture all year long.

It's not surprising that under these conditions are found the most exuberant plant formations, with the richest variety of species on the planet: the tropical rain forests.

THE PREDOMINANCE OF TREES

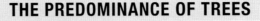

One of the characteristics of the tropical jungle is the predominance of trees. These trees also have a very different appearance from the trees of other types of forest. Their trunk is very straight and slender in comparison with their tremendous height. The bark is smooth and light in color. The crown is rather small. These trees have very superficial roots, so the tallest trees commonly have the typical props at their base. The lowest branches are found at a significant height above ground.

Many animals that live in the tropical jungle, such as sloths and certain monkeys, never touch the ground even once during their lifetimes.

Comparison of the outer appearance of trees from different types of forest.

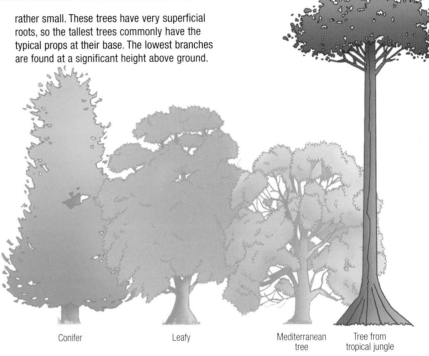

Conifer   Leafy   Mediterranean tree   Tree from tropical jungle

The foot of the trunk of a tree in a tropical jungle with the typical buttressing.

Humans have eliminated many areas of the tropical jungle to harvest woods with great commercial value, such as ebony, mahogany, and banyan.

FRUITS IN STRANGE PLACES

The flowers and fruits of the small and medium shrubs and trees in the jungle often sprout in very surprising locations, such as tree trunks, thick leaves, and short, leafless stems.

The cocoa tree. Powdered cocoa is obtained from the seeds of its fruit.

JUNGLE LEAVES

The leaves of the jungle are of the laurel type, but they end in a characteristic turned-down **dripping** tip. Also, newly sprouted leaves are not green, since they have no chlorophyll; rather, they are red, crimson, light purple, or even white, and they hang down as if they were wilted. The first time you walk into a jungle it's easy to "see" flowers where there are merely new leaves.

Typical branch of a tree in the tropical jungle.

DIFFERENT TYPES OF LEAF

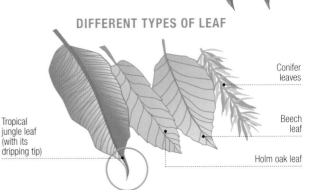

Tropical jungle leaf (with its dripping tip)

Conifer leaves

Beech leaf

Holm oak leaf

Among the lush vegetation of the tropical jungle, vines seek the light and the ability to develop; sometimes they strangle the trees they use for support.

Introduction

Plant Anatomy

Plant Physiology

Reproduction

Flower, Fruit, and Seed

Ecology and Evolution

Algae

Fungi

Plants

Plants with Flowers and Fruits

Plants and Their Environment

Aquatic Plants

Wild Plants

Domesticated Plants

Gardens

Alphabetical Subject Index

VINES

Vines are the climbing stems of many plants that have opted to reach the light quickly without the great investment involved in growing a trunk. That way they can climb among the crown of the trees and use them for support. There also are vines in other types of forest, but they are very highly developed in jungles.

In tropical jungles, there are **vines** as thick as a person's arm and some of them are almost 750 feet (240 m) long.

The wild pigs of the jungle are scavengers. They eat the corpses of monkeys and other animals that fall from the trees when they die.

Staghorn is an epiphytic plant that gets its name from its resemblance to the antlers of certain cervids.

FREEDOM FROM THE SOIL

Many small plants use another strategy to reach the light; they free themselves from the soil and live on the trunks and branches of trees. As with the vines, this is not a case of parasitism, since they merely use the tree for support. These types of plant, known as **epiphytes**, include mosses, ferns, and beautiful flowering plants. Each of these plants uses different tricks to provide itself with "soil," humus, and water in the hollows, cracks, forks, and folds of the trunks and branches.

A CLOSED NUTRIENT CYCLE

Tons of plant remains fall to the jungle floor every day. Shortly after a storm, the water that drains down from leaf to leaf inundates the soil, where ants, termites, and other consumers of detritus begin the decomposition process. Given the high temperature of the environment and the multitude of microorganisms present in the soil, the mantle is transformed almost automatically into minerals that plants can assimilate. They absorb them, so they get by fine with their superficial roots. This is a closed and quick cycle.

There is a tremendous activity of transformation that takes place on the ground but is hardly visible.

Because teak is so durable, it is used for making outdoor furniture.

TEAK

Many tables and benches intended to be kept outdoors are made from wood that comes from a tree known as the **teak**, which doesn't deteriorate with moisture. These trees form forests in tropical areas where it rains a lot but not for the whole year, as in the jungle. These forests are called **deciduous tropical forests** because many of their trees lose their leaves during the dry season.

AQUATIC PLANTS

To speak correctly, we should make a distinction between algae and aquatic plants. Algae are the aquatic plants *par excellence*, but they are not real plants, because they don't have tissues, roots, stems, or leaves. However, there are plants that live in the water, either completely or partly submerged, whether floating freely or merely letting their leaves float. Many of them are plants that bear flowers, that is, land plants that have become adapted to aquatic life.

SOAKING THEIR FEET

Many plants colonize soils that are saturated or even submerged to a depth of a little more than three feet (1 m). They live with their roots or rhizomes attached to the bottom but with their stems and leaves outside the water. To keep their roots from smothering for lack of oxygen, these plants have leaves filled with air that they send to the roots.

The beauty of some species of water lily makes them popular as decorative plants in ponds and some gardens.

Reeds (left) and cattails (right) are freshwater, emerged plants.

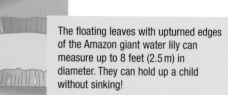

Amazon giant water lily.

FLOATING LEAVES

Other plants live anchored to the submerged bottoms by means of their roots and keep their leaves floating on the surface. These leaves commonly are robust and circular in shape, and they have a very flexible petiole that's longer than the depth in which the plant lives. That way they are safe from the damage that the water's movement would cause.

The floating leaves with upturned edges of the Amazon giant water lily can measure up to 8 feet (2.5 m) in diameter. They can hold up a child without sinking!

DEFORMED TREES AND FOUL-SMELLING BUBBLES

This is what you see at first glance in a mangrove stand, if your vision isn't obstructed by a cloud of insects or other little creatures that add nastiness to these locations. Mangrove stands are partially submerged forests that form on low, tropical coasts and are subject to tides. A mass of tangled roots like pillars raises the base of the trees above the rotten quagmire.

Other aerial, respiratory roots grow upward from them. There is a lot of life in a mangrove stand and lots of fish nearby.

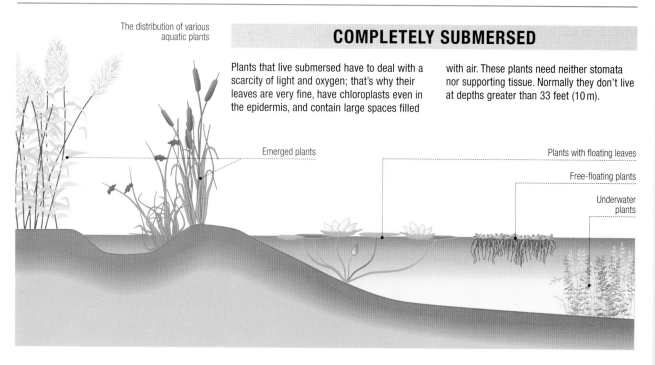

The distribution of various aquatic plants

Emerged plants

Plants with floating leaves

Free-floating plants

Underwater plants

COMPLETELY SUBMERSED

Plants that live submersed have to deal with a scarcity of light and oxygen; that's why their leaves are very fine, have chloroplasts even in the epidermis, and contain large spaces filled with air. These plants need neither stomata nor supporting tissue. Normally they don't live at depths greater than 33 feet (10 m).

Certain plants, such as *Eichornia* (bottom) manage to slow down navigation in some American rivers. At the top, a *Lemna*, a type of floating plant that forms a green carpet on the water, especially in backwaters.

FREE IN THE WATER

There are plants that live at the mercy of the water, either on the surface or submerged, without setting down roots in any type of substrate. Some of the latter have leaves that are converted into traps to capture tiny animals that also live at the mercy of the water (zooplankton), that way they make up for the lack of mineral nutrients dissolved in the water.

Submersed plants

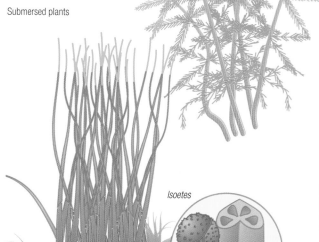

Myriophyllum

Isoetes

UNDERWATER MEADOWS

These dry "algae" in the shape of a ribbon that the waves cast up on the beach are not algae. They are **flowering plants** known as **posidonia**, which are adapted to life on the well-lighted bottom of the ocean near the coast, where they form vast meadows that are real cauldrons boiling with life. They have a rhizome as thick as a finger, with roots, at the end of which the bundles of leaves are located. Fortunately, few marine animals eat them, but urchins devour them day and night. Still, their worst enemy is the dragnets of fishing boats.

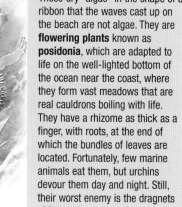

A sea urchin feeding on posidonia.

EDIBLE WILD PLANTS

Before practicing agriculture, humans lived by hunting and collecting plants or plant parts. But you mustn't think that this practice belongs only to prehistory. There are still people and tribes that feed themselves, if not entirely, at least mainly in that way. Even in countries that base their food production on agriculture and the raising of livestock, wild plants are still eaten, many of which achieve major commercial value, such as certain fungi.

FAVORITE MUSHROOMS

The same mushrooms are not found in all areas, nor are they equally esteemed in all places. The Mediterranean area is one of the regions that most appreciates the consumption of wild mushrooms. There are a great number of edible species that grow in the woods, but, in practice, most consumers gather and eat only the tastiest ones.

The morel (*Morchella vulgaris*) is edible, and it is found among elms and ash trees in the spring.

Lepista nuda

TENDER LEAVES AND STEMS

Even though many wild plants that people gather have counterparts among cultivated plants, some have a stronger taste, and that's why they are highly esteemed. One example is asparagus. Even though the cultivated ones are thicker and fleshy, the wild asparagus are much tastier. But that's not the case with all plants; wild chestnuts, for example, are of lower quality than the cultivated ones.

Asparagus shoots are new stems that come up every year at the end of winter. If they aren't cut, they turn into stems with leaves. They can be cooked in many ways.

MAGUEY

When the maguey plant blooms, it produces a long stem, at the end of which the flowers appear. If you cut this stem, it exudes a liquid known as hydromel; in Mexico it is turned into an alcoholic drink called **pulque.**

Gentian is used in making many aperitifs.

WILD FRUITS

Nowadays, these are the most highly valued products of nature, along with mushrooms, both for eating fresh and for making ice creams, jellies, and desserts. They also are used in modern cooking to create original bittersweet flavors.

Animals that live in the wild don't ingest plants that may be toxic to them. Their instinct guides them.

DRY WILD FRUITS

There are many cultivated dry fruits that also exist in the wild, where they often are smaller and tastier. Others are not cultivated, but if they are forest species, sometimes they are used for replanting, so they provide an additional benefit. Some pine plantations, for example, have piñon pines, and when the pines are harvested, the pine nuts also can be sold.

The pine nuts of the piñon tree are not really fruits but rather seeds. The Brazil nut is the seed of a very common tree in the Amazon jungle; it is sold throughout the world.

Indian bread is known by that name because the natives of South America ate it at the time the Spanish arrived.

FOR LIVESTOCK

We humans consume wild plants in more ways than by eating them directly. A large part of the animal food that we eat has its origin in **natural feeds** and other wild plants that the livestock eats. In the Eastern United States, many wildlife species prized by hunters feed on the acorns of scrub oaks, including turkey, deer, bear, grouse, and quail. In Mexico, the fruits of the red oak are an important food source for livestock.

The blackberry is a bushy, spiny plant that is used for edging along fields and roads. The fruit can be eaten raw or used to make delicious jams.

The mongongo nut makes up half the vegetarian diet of the bush people of the Kalahari desert.

The flavors of many ice creams come from wild fruits.

Flower and fruit of the blackberry; it ripens at the end of summer.

83

MEDICINAL PLANTS

Primitive humans used certain plants as remedies for pains, and there are still people who base their medical practice exclusively on the use of plants with healing properties. In industrialized nations, pharmaceutical products are used more commonly, but many of these substances are plant extracts. Other drugs are manufactured in pharmaceutical laboratories and industries by "copying" the chemical composition of the **active ingredient** of the plants that had curative effects.

WHAT IS A MEDICINAL PLANT?

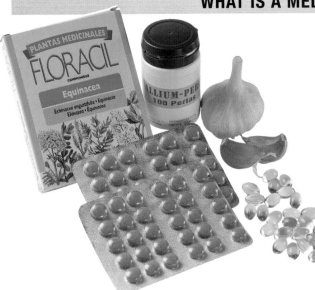

Many plants contain chemical substances that can produce special effects on the body of a living creature, whether beneficial or harmful. These substances are known as the **active ingredients**. Medicinal plants are those that have one or more active ingredients that are capable of preventing, alleviating, or curing illnesses. The effects produced by the active ingredient of a medicinal plant are closely linked to **how the plant is used** and the quantity that is applied or administered, in other words, the **dosage**.

Many drugs and remedies are made from the active ingredients contained in medicinal plants.

Herbalist's shops sell medicinal plants and prepare formulas for traditional remedies.

CURATIVE SUBSTANCES

The **active ingredients** of a medicinal plant are its most important feature. These ingredients usually are concentrated in a specific section of the plant, such as the root, the leaves, the flowers, and so on.

The important part of the chamomile plant is the heads or inflorescences. Herbal tea made from chamomile has digestive properties.

THE ORIGIN OF ASPIRIN

People had been using an extract from willow bark for centuries to alleviate pain, and this product was sold in pharmacies in the time of your great-grandparents. Then chemists succeeded in isolating the active ingredient of this extract and synthesizing it in the laboratory; that is, they discovered how to produce it artificially through chemical synthesis. It was marketed under the name of aspirin.

THE OPIUM PLANT

In your grandparents' time, almost all country families had one or two **poppy** plants in a corner of their yard. When anyone had a toothache, they drank a tea made from this plant. The poppy is an example of a plant that has **alkaloids**, which are substances with a much stronger action than normal active ingredients. The poppy concentrates these substances in its fruits, which are the source for **opium** and **morphine**, two powerful, soothing narcotics.

It is not easy to distinguish between the plain poppy and the opium poppy.

GATHERING AND STORING

Medicinal plants or parts of them are gathered at the moment when the content of active ingredients is at the optimal level. Then they are normally put aside to dry in different ways: they are hung up, spread out on papers or racks, or dried in an oven. This last technique is used especially for drying fruits. Barks and roots are dried after they are chipped up. Finally, the dried and crushed plants, which have been pulverized in a mortar, are kept in containers to keep them until needed for making a **medicinal tea**.

WAYS TO PREPARE A MEDICINAL TEA

TYPE OF TEA	PREPARATION
Infusion	Pour boiling water over the medicinal plant and cover the cup.
Cooking	Boil the medicinal plant in a closed container for ten to twenty minutes.
Steeping	Let the medicinal plant soak for several hours.

→ **Syrups** are made by adding the medicinal plant to a solution of sugar and water.

Medicinal tea made by infusion.

To boil up a medicinal tea, strain or filter the liquid after cooking the medicinal plant for the appropriate time.

To prepare a medicinal tea by steeping, pour hot water over the medicinal plant, and let it soak for the prescribed time.

A FATAL DOSAGE

Never use medicinal plants or herbal supplements without the help of a physician. The chemical substances that the plants contain may be beneficial in small doses, but they can be very toxic in doses that appear normal. **Belladonna** is one example. Its active ingredients and alkaloids are very useful in medicine, but people who have ingested the fruits of the plant through ignorance have died from respiratory paralysis.

Belladonna

The water in which the leaves of nettles are boiled, applied to the scalp, slows down hair loss. The juice of the chopped leaves is rubbed in to prevent dandruff.

WHAT ARE THEY USED FOR?

Normally, a single plant produces more than one beneficial effect, but it tends to stand out as a remedy for a specific type of ailment. Similarly, different plants can be used for the same ailment. At the herbalist's, they are already classified according to their healing effects, and they even have mixtures in the right proportions to produce the desired effects.

Valerian is a medicinal plant that yields extracts effective in treating anxiety, insomnia, gastrointestinal spasms, and muscle spasms.

AROMATIC PLANTS

There are plants that give off an aroma or special perfume that's noticeable even without touching them. You may have noticed that certain plants emit their fragrance when you brush them and that others don't release their perfume unless you mash them with your fingers. All these plants owe their fragrance to the essences that they contain in their tissues. These **essences** or **essential oils** are chemical substances that are soluble in alcohols and oil, and can be extracted from the plant and used in manufacturing all-natural lotions, colognes, perfumes, and aromatics.

FLOWERS WITH TWO LIPS

One of the plant families with many aromatic species is the **labiates.** You can use a simple magnifying glass to see that the corolla of their flowers consists of an upper lip formed by two petals and a lower one made up of three. The aromatic labiates are used in perfumes and cosmetics, as natural aromatics in foods, and also as medicinal plants. They have concentrated essences in tiny vesicles present in the hairs of the leaves and flowers.

An essence is a liquid with a high concentration of aromatic substances.

These three essences, lavender, eucalyptus, and rosemary, are typical.

SOME AROMATIC LABIATE FLOWERS

Wild thyme

Wild chervil is a representative umbellifer. Its leaves are used for adding flavor to sauces and salads.

A PARASOL OF FLOWERS

Another family of plants known as the **umbellifers** is likewise rich in aromatic species. These are characterized by their clusters of flowers in the shape of an open parasol. They have tiny channels filled with essences on almost all parts of their body but especially on the fruits, which are so small that they look like seeds. Another characteristic of the umbellifers is the repeated segmentation and the nearly omnipresent highly developed veining of their leaves.

Common mint

Lavender

HEMLOCK

The name hemlock is associated with poisoning because it truly is a poisonous plant, although it's also used medically because of its anesthetic power. This umbellifer is very common in the wild and even grows at the edges of roads. You shouldn't touch it, for its alkaloids can enter the body through the skin.

A FLOWER COMPOSED OF HUNDREDS OF FLOWERS

If you use a magnifying glass to look at a daisy or a chamomile plant, you will see that it is composed of a large number of tiny, compressed flowers: What appears to be petals are **ligules**, that is, scalelike projections of the flowers. Plants with this type of inflorescence make up the family of **composites**, which also contains many aromatic species.

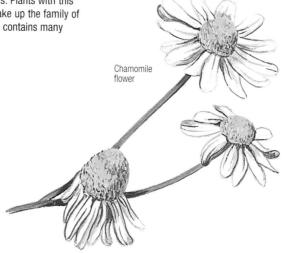

Chamomile flower

Rock tea grows in cracks in crags, especially in the Pyrenees Mountains; it blooms from June through August. The caps are gathered just before they open, and, once they are dry, they are used for making infusions that people drink just like regular tea.

SPECIAL FLOWERS

What commonly is called cologne is in reality perfumed water with an essence dissolved in a little alcohol. A perfume, in contrast, is a concentrated essence. There are aromatic plants whose flower has a unique perfume that is present only in the flower, such as the rose, the jasmine, the violet, and the orange blossom.

Along with its natural beauty, the rose commonly gives off a pleasant scent.

Orange and lemon trees, both wild and cultivated, produce blossoms.

VANILLA AND CHOCOLATE

When the Spanish conquered Mexico, they discovered that the natives flavored chocolate with some aromatic powders obtained from the fruit of a plant. This plant, the vanilla plant, is a climbing orchid that grows only in tropical climates. The aromatic essence is only in the unripe fruit, so it must be harvested green and then dried in a very controlled way. Vanilla is used widely in confections, desserts, the manufacture of chocolate, and the distilling of cognac and rum.

Nowadays, much of the vanilla and other flavors that are used are manufactured synthetically.

PLANTS FOR PRODUCTION

The foods of plant origin that we buy in shops and markets, such as rice, potatoes, lettuce, and oranges, are grown by farmers. Cultivated plants originated from wild plants that humans have modified over thousands of years to facilitate harvest, improve the edible parts, and improve quality. But farmers also cultivate plants that are used in making many things, such as clothing, baskets, tobacco, fiber sandals, and colors for dying fabric.

THE DOMESTICATION OF PLANTS

Primitive humans domesticated plants, just as they tamed animals. They always kept the seeds of the best plants or the most vigorous and productive hybrids. After many years, the cultivated plant hardly resembled the original wild one and was better adapted to the conditions created by humans. The more domesticated a plant is, the more dependent it is on the care of the farmer, because it has lost the characteristics that protect it from herbivores, such as thorns and toxic substances, as well as the ability to reproduce by dissemination.

Corn is an example of a highly domesticated plant that is practically incapable of reproducing without the help of humans. If a cornfield is abandoned, when the ears fall to the ground, the kernels cannot be buried to reproduce new corn plants by themselves.

The **ear of corn** that is grown nowadays is seven or eight times longer than the ones on the first plants of this grain that were domesticated.

FOOD OILS

Have you ever wondered why **olive oil** is much tastier than the other oils in the store, such as **soy, corn, sunflower,** and **peanut?** Olive oil comes from a fruit, the olives, which are simply squeezed. In contrast, the other oils are obtained from the plants' seeds; first the oil is extracted from the seeds with a chemical solvent and then it has to be refined by using further chemical processes.

BASKETS, RUGS, HATS . . .

The fibrous stalks of plants are very tough and often flexible. Baskets are made from **strips of chestnut** or **birch bark,** and **wicker** (young willow branches) is used for many things, including furniture making. **Sweet grass** is, especially by Native Americans, used for making baskets and hats. **Straw** is used in making hats; baskets, chair seats, and rugs are made from palmetto leaves. Countless objects also are manufactured by using **cane, reeds,** and **rushes.**

PLANTING TO PRODUCE SEEDS

The most important crops of this type are **wheat**, **rice**, and **corn**, which are the food base for most humans. Other cultivated seeds include **millet, sorghum, oats, barley,** and **rye,** which also are cereals and legumes such as **beans, lentils, chickpeas, broad beans,** and **peas.**

Wheat is an important food for humans. The grain is ground and the resulting flour is used to make bread, crackers, and pasta (macaroni, spaghetti, etc.).

Legumes, especially beans, are an important source of protein for many people who lack the means to eat meat regularly.

ROOTS

In the tropics, a lot of roots rich in starches are grown; these include **sweet potatoes, tapioca, taro,** and **yams.** In temperate climates, **turnips**, **carrots**, and **sugar beets** (for producing sugar) are grown.

The avocado has been cultivated for about 7,000 years in Central America, where it is called "the butter of the poor." The same plants are the basis of modern production.

COFFEE, TEA, WINE, COCOA

Except for **wine**, which is made by fermenting **grapes**, which grow in Mediterranean climates, most of the plants that are grown for making drinks are from tropical countries, such as **coffee, tea,** and **cocoa.**

Cinnamon is the bark from the branches of the cinnamon plant; it is dried, and the epidermis is removed.

IN SEARCH OF RARE PLANTS

Some 600 years ago, Europeans adventured across the seas in search of plant products that could improve the taste of foods, such as pepper, ginger, cloves, and cinnamon. These are **spices**. At that time, there were spices that were more valuable than gold, and merchants kept their places of origin a secret. Later on, spices were grown on a large scale, and they ceased to be exotic products.

Coffee is made from the toasted seeds of the coffee plant, which is a shrub.

FABRICS FROM NATURAL FIBERS

Synthetic fibers, such as polyester, didn't exist 60 or 70 years ago. Clothing was made either from sheep's wool or from **cotton** or **linen**, two fibers from plants that were grown on a large scale not long ago and that are still cultivated but on a smaller scale. **Hemp** fibers are used to make ropes, and **jute** fibers are used for bags and rugs.

Cotton fiber, which is used to make many kinds of soft fabrics, comes from the epidermal cells of the cottonseeds.

In tropical countries, sugar is extracted from sugar cane; in temperate climates, sugar comes from the swollen root of the **sugar beet**. In both cases, the sugar is produced from the juice.

ARBORICULTURE

Among the first trees domesticated by humans were the **olive**, the **palm**, and the **avocado**. Nowadays, countless species and varieties of fruit trees, both tropical and temperate, are cultivated. The cultivation of fruit trees is called **arboriculture**.

The wild species that could be the ancestors of tobacco contain too much nicotine in their leaves. The first tobacco farmers had to create hybrids with a lower nicotine content.

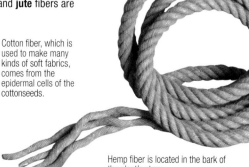

Hemp fiber is located in the bark of the plant's stem.

PLANTS FOR DECORATING THE HOUSE

If you like plants, you can brighten up the interior of your house, as long as you take care to know the needs each one has with respect to light, moisture, irrigation, type of soil, mineral nutrients, and temperature. Remember that **houseplants** need exactly the same living conditions as their wild relatives, adapted to the environmental conditions of the place where they grow. So they will thrive indoors only if they find conditions (light, temperature, etc.) similar to those that are found in the wild.

INDOOR HEROES

If you don't know where a plant originated and the conditions under which it lives in the wild, you can get some clues about how to treat it from the way it grows, the type of leaf, and other characteristics. The most resistant plants are those that have rigid, leathery, long-lasting leaves and grow slowly. These include **palms**, **philodendrons**, **ficuses**, **dracaenas**, **aspidistra**, **sansevieria**, and **green iris**, among others.

The ficus has large, brilliant green leaves. They shine more if they are wiped clean periodically.

Some plants, such as camellias, "take great offense" when they are moved or turned while their floral buds are developing and opening.

LIGHT

In general, flowering plants require more light than nonflowering ones. Plants commonly show signs of inadequate light by turning a paler color.

GIVE IT A PLACE IT LIKES

When you look for a place for your plant, it's very important to consider the availability of light and where the direct sunlight reaches. You also have to remember that in a closed area the temperature and dryness of the air increase with height, so you have to put the less sensitive plants that need less water in the higher places.

The pothus or scindapsus doesn't need much soil, but it has to have plenty of water, and it grows quite quickly.

SHADE AND MOISTURE

Smooth leaves and soft, juicy stems are signs of plants that live in moist, shady soils. These are the plants that cover the floor of damp, tropical jungles or moist forests in warm climates, such as leafy begonias, gloxinia, African violets, ferns, and mosses.

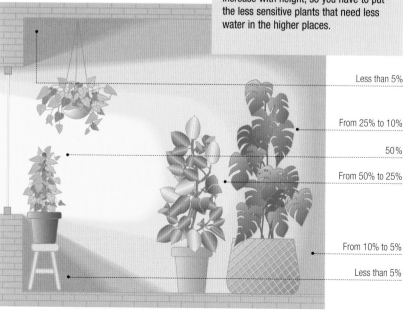

Less than 5%

From 25% to 10%

50%

From 50% to 25%

From 10% to 5%

Less than 5%

The petunia is a plant from South America; there are several wild varieties. The hybrid varieties are used as a decorative plant because the flowers are very attractive.

When you buy a houseplant, you need to find out about how much light and moisture it needs. The drawing shows the average light that a plant receives depending on its location.

JUNGLE EPIPHYTES IN THE HOME

The **epiphytic plants of the tropical rain forest** live on other plants, and their "roots" are reduced to small hooks. It's not easy to imitate in the home the conditions to which these plants are adapted. They like humidity in the air, so the dryness that home heating produces is not good for them. And if you spray the area, you have to keep the water from falling onto the leaves.

This is the case with the **dendrobium orchid**, **Christmas cactus**, and the **staghorn fern**.

Orchids need very specific care, depending on the variety.

ORIGINAL PLANTS

Because we are used to the uniform green leaves of most plants, we are attracted to the varieties that have leaves with stripes or yellow dots. These plants, which are referred to as **variegated**, owe their originality to the partial or total lack of green pigment in their leaves. They are very pretty, but since they are characterized by a lack of pigment, they also are more delicate. They bloom less frequently or not at all, and they grow more slowly.

The cacti that are planted in a pot or a corner of the garden don't need much water.

LOTS OF LIGHT

Plants that develop branches quickly and continually form lots of young leaves usually require lots of light. However, don't forget that the conditions next to a window in strong, direct sunlight inside your house never occur in the wild; the warming produced by the glass panes can burn the plant.

If you suddenly move a plant from a dark location to a bright one it may experience damage from excessive light; the same can happen if the light varies in the place where the plant is kept.

The coleus is very attractive because of the coloration of the leaves.

Too much light, and especially too much heat, can harm plants. It's a good idea to reduce the excess exposure with a grate or an awning.

ALL-TERRAIN PLANTS

Don't be misled into thinking that just because the **succulent plants** and the **cacti** are less delicate they lack beauty. Since they are plants that have adapted to desert conditions, they normally need a period of **dry rest** in the winter.

The blooming of certain succulent plants is very attractive, and it commonly lasts for many weeks.

The Vriesea has elongated leaves in which the central part takes on a very bright red color.

GARDENS

If you look at a garden, you quickly see that it's not just a series of plants, nor a collection of pretty plants. The person who conceived it arranged the plants in a certain order, imagining how people would feel in the garden once the trees had grown and the various plants had formed different green masses of varying shades. The garden also may have been planned so there would be a place for games, shade to sit under and chat on summer afternoons, and even an area to sit and meditate.

GARDEN TREES

The trees in a garden serve different functions. The most decorative ones are the **conifers**, which keep up their beautiful green foliage all year long, and the exotic **palms**. The so-called **flowering trees** are included for the beauty of their flowers at a certain time of year; these include the **almond**, the **cherry**, and **crabapples**. Deciduous trees with a rounded crown are used to create shade in the summer in areas where sun is wanted in the wintertime. These are the **shade trees**, such as red maples, ash, American sycamore, and river birch.

The London plane is one of the most commonly used trees in cities because it offers pleasant protection against the sun during the summer, and because it's a deciduous tree, it lets through the weak sunlight in the winter. The photo shows the leaves and the fruit.

HOW DO YOU PLANT A TREE?

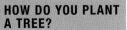

Several days before buying the tree you should dig a hole about three feet (1 m) deep and three feet across. Mix the dirt from the hole with manure or compost. When you plant the tree, first pour water into the hole and add some of the dirt; place the roots of the plant on that and add the rest of the dirt, pressing it against the roots. Finally, water from above so that the water filters down and reaches all the roots.

There are many varieties of palm trees, some of which are highly valued for their fruits (such as the coconut tree and the date palm). But they also are used as decorative trees because of their slenderness. To keep them growing well, the lower leaves are periodically cut off the crest, leaving the trunk visible (as in the photo).

MINIATURE GARDENS

The art of growing dwarf plants has the advantage of not requiring large areas. This involves growing trees and shrubs in pots by using techniques that interfere with the normal development of the roots and stems. These miniature trees, which are known by the term **bonsai,** live just like their full-size colleagues.

CLUSTERS

The beauty of a garden comes not from individual plants, but rather from masses of plants of a single species or a few similar species. These masses can be green or flowering. Shrubs are great for making up clusters.

Although it's hard to believe, the area occupied by this garden was a huge cement quarry a few decades ago. This is the famous Butchart Gardens in Canada.

HEDGES

Hedges are green barriers that are used to mark the boundaries of a yard or to separate spaces. The most beautiful hedges are made up of trees, shrubs, and plants of different heights and shapes, in imitation of natural vegetation. But there also are hedges composed of a single species; they are commonly trimmed every year, sometimes to form geometrical patterns.

In addition to their esthetic function, hedges mark off spaces in a yard or a property, protect other plants and flowers from the effects of wind, act as a noise and pollution barrier, and can even create amusing mazes.

GARDENS FOR MEDITATION

In Japan, most gardens are planned to provide peace and inner tranquility for the people who use them. These are the so-called **Zen gardens**. The placing and arrangement of the features that go into the gardens, such as stones, water, and plants, as well as the designs on the gravel walkways, promote meditation and harmony with nature.

GARDENS FOR TRAVELING

If you walk through a botanical garden, you will see plants that come from all areas of the world. Some of these plants are very rare or are from sites that are very difficult to reach. One way to take a trip around the world is to walk through one of these gardens.

LAWNS

Lawns are a mixture of various gramineae and other grasses adapted to being cut to ground level and walked on. They are a beautiful natural carpet but in areas where the summer is long and hot, they require a lot of watering. That's why many Mediterranean gardens are made without a lawn or with a very small one. Broad expanses of lawn are appropriate for humid climates without too much sun, such as in the gardens of the Cantabrian coast, Great Britain, and the north of France.

ROCK GARDENS

Rock gardens are islands of species appropriate to arid terrain planted among appropriately piled-up rocks with the right additions of soil. They serve a decorative function and they give a wild touch to the garden. Plants for rock gardens are adapted to drought and strong sun, and they insinuate their roots among the rocks.

French gardens (e.g., the Schönbrunn Gardens in Vienna) consist of large areas with bunches of flowers carefully arranged and no intermediate elements to interfere with vision across large spaces.

A lawn is one characteristic of English gardens. The photograph shows the gardens of Hampton Court, near London.

ALPHABETICAL SUBJECT INDEX

Plant
Anatomy

Plant
Physiology

Reproduction

Flower, Fruit,
and Seed

Ecology and
Evolution

Algae

Fungi

Plants

Plants with
Flowers and
Fruits

Plants and Their
Environment

Aquatic Plants

Wild Plants

Domesticated
Plants

Gardens

**Alphabetical
Subject Index**